GUIDELINES FOR THE MANAGEMENT OF CHANGE FOR PROCESS SAFETY

This book is one in a series of process safety guideline and concept books published by the Center for Chemical Process Safety (CCPS). Please go to www.wiley.com/go/ccps for a full list of titles in this series.

GUIDELINES FOR THE MANAGEMENT OF CHANGE FOR PROCESS SAFETY

Center for Chemical Process Safety
New York, New York

It is sincerely hoped that the information presented in this document will lead to an even more impressive safety record for the entire industry. However, neither the American Institute of Chemical Engineers, its consultants, CCPS Technical Steering Committee and Subcommittee members, their employers, their employers' officers and directors, nor ABSG Consulting Inc. and its employees warrant or represent, expressly or by implication, the correctness or accuracy of the content of the information presented in this document. As between (1) American Institute of Chemical Engineers, its consultants, CCPS Technical Steering Committee and Subcommittee members, their employers, their employers' officers and directors, and ABSG Consulting Inc. and its employees and (2) the user of this document, the user accepts any legal liability or responsibility whatsoever for the consequence of its use or misuse.

A Joint Publication of the Center for Chemical Process Safety of the American Institute of Chemical Engineers and John Wiley & Sons, Inc.

Published by John Wiley & Sons, Inc., Hoboken, New Jersey.
Published simultaneously in Canada.

Library of Congress Cataloging-in-Publication Data is available.

ISBN 978-0-470-04309-7

Printed in the United States of America.

10 9 8 7 6 5 4 3 2 1

Guidelines for Management of Change for Process Safety
is dedicated to the memory of

Sanford Schreiber
1925 – 2007

Sandy Schreiber joined CCPS in 1986 after a 28-year career with Allied-Signal, where he was Director of Corporate Safety and Loss Prevention. Sandy's early development of the twelve technical elements that must be part of any chemical process safety management program was a ground-breaking concept. These elements were published in *A Challenge to Commitment* and set CCPS on a path of influence and success. The subsequent development of the 4-volume series of process safety management guidelines, leading to the recent book, *Guidelines for Risk Based Process Safety*, owes a debt to Sandy's vision.

We have lost a good friend, and the industry has lost a pioneer.

CONTENTS

PREFACE

The American Institute of Chemical Engineers (AIChE) has been closely involved with process safety and loss control issues in the chemical and allied industries for more than four decades. Through its strong ties with process designers, constructors, operators, safety professionals, and members of academia, AIChE has enhanced communications and fostered continuous improvement of the industry's high safety standards. AIChE publications and symposia have become information resources for those devoted to process safety and environmental protection.

AIChE created the Center for Chemical Process Safety (CCPS) in 1985 after the chemical disasters in Mexico City, Mexico, and Bhopal, India. The CCPS is chartered with developing and disseminating technical information for use in the prevention of major chemical accidents. The center is supported by more than 100 sponsors within the chemical process industry who provide the necessary funding and professional guidance to its technical committees. The major product of CCPS activities has been a series of guidelines to assist those implementing various elements of a process safety and risk management system. This book is part of that series.

Uncontrolled changes have directly caused or contributed to many major accidents that have occurred within the chemical process industry and allied industries. Many industries and companies recognize the importance of careful management of change (MOC) for ensuring the safety of process operations and the quality of manufactured goods. The concept and the need to properly manage change are not new; many companies have implemented MOC systems. Yet incidents and near misses attributable to inadequate MOC systems, or to subtle, previously unrecognized sources of change (e.g., organizational changes), continue to occur. To improve the performance of MOC systems throughout industry, managers need advice on how to better institutionalize MOC systems within their companies and facilities and to adapt such systems to managing non-traditional sources of change. CCPS is helping to fulfill this need through the publication of these guidelines.

The purpose of this book is to define the important features of MOC systems. MOC systems help ensure that changes to the design, operation, maintenance, and organization of facilities will not adversely affect employees, the public, or the environment. MOC systems are used not only for process safety purposes, but also to manage quality, security, environmental, and organizational risk issues. This document outlines a process that can be used for designing, developing, installing, operating, maintaining, and improving MOC systems at individual company sites and at corporate or support locations. The appendices contain examples, flowcharts, and forms that should be useful to personnel who are implementing new MOC systems or improving existing ones. The enclosed CD contains an MOC system design tool, an MOC system diagnostic tool, and examples of typical MOC system procedures and forms.

The hazards associated with a proposed change are not limited by the size or complexity of the facility in which the proposed change is to be implemented. Thus, just because a facility may be small or have relatively simple processes (e.g., storage and unloading), the need to properly manage change is no less important than at larger or more complex facilities. Also, managing change at small facilities is not necessarily easier than implementing an MOC system at a large facility. Each situation carries its own special challenges. Large facilities, where making adjustments to the facility culture is often more difficult, can find that gaining consensus on the procedures for managing change is equally difficult. Smaller facilities, which are often more receptive to change, may lack the resources (e.g., people, technical specialties) that are more common at large companies/facilities. To help meet the needs of smaller facilities, this book includes an overview of the MOC chapter from the CCPS book entitled *Guidelines for Risk Based Process Safety*, which promotes the efficient design, implementation, and improvement of "just fit-for-duty" management systems, including MOC.

This book is intended for an audience ranging from facility and corporate managers of process safety to workers who have differing levels of knowledge about the principles of safely managing change. This book is primarily designed to equip people responsible for MOC systems with new ideas for implementing and improving MOC systems. However, it may also be used as a training aid for companies teaching process safety management and MOC concepts to new employees.

ACKNOWLEDGMENTS

The American Institute of Chemical Engineers (AIChE) and its Center for Chemical Process Safety (CCPS) express their gratitude to all of the members of the Management of Change (MOC) subcommittee and their CCPS member companies for their generous efforts and technical contributions in the preparation of these guidelines.

The CCPS MOC committee was chaired by Jim Muoio (Lyondell). The committee included (in alphabetical order): Susan Behr (Sunoco), David Cummings (Du Pont), Tom Dileo (Albemarle), Robert Dupree (Basell), Christy Franklin (RRS Engineering), Wayne Garland (Eastman), Bill Lash (BP), Steve Marwitz (Formosa Plastics), Mike Moriarty (Akzo Nobel), Lisa Morrison (BP), Jeffrey Philliph (Monsanto), Mike Rogers (Syncrude), Tony Santay (Air Products), Dan Wiff (Nova Chemicals), and Gary York (retired, formerly U.S. Rhodia). The CCPS MOC subcommittee Staff Consultant was Bob Ormsby. In addition, Appendix D on electronic MOC applications was based on materials provided by MOC subcommittee members Wayne Garland and Mike Rogers and peer reviewer David Drerup.

CCPS wishes to acknowledge Steve Arendt and Walt Frank of ABS Consulting who were the principle authors of this book. Bill Bradshaw and Jim Thompson of ABS Consulting also contributed to these guidelines, along with Leslie Adair and Karen Taylor, editors; Scott Campbell and Paul Olsen, graphic designers; and Susan Hagemeyer, word processor.

We thank the following people and their organizations who generously contributed their time and expertise in providing a peer review of the book:

John Alderman, *RRS Engineering*
Luigi Borriello, *PPG Europe*
Gary Carrithers, *Rohm & Haas*
Don Connolley, *BP*
Susan Cowher, *ISP Technologies*
David Drerup, *Data Systems & Solutions*
Bob Gale, *Emerson*
Greg Keeports, *Rohm & Haas*
Pete Lodal, *Tennessee Eastman*
Jack McCavit, *JLM Consulting*
Henry Ozog, *ioMosaic*
Adrian Sepeda, *CCPS Emeritus*
John Wincek, *Croda*

ITEMS ON THE CD ACCOMPANYING THESE GUIDELINES

MOC System Design Tool (Excel Spreadsheet)

MOC System Diagnostic Tool (Excel Spreadsheet)

Example MOC System Procedure and Forms

LIST OF TABLES

LIST OF FIGURES

ACRONYMS AND ABBREVIATIONS

ACC	American Chemistry Council
AIChE	American Institute of Chemical Engineers
ASP	application service provider
CCPS	Center for Chemical Process Safety
CFR	*Code of Federal Regulations*
COMAH	Control of Major Accident Hazards
DCS	distributed control system
EDMS	electronic data management system
eMOC	electronic management of change
EPA	Environmental Protection Agency
FIBC	flexible intermediate bulk container
HAZOP	hazard and operability
ISO	International Organization for Standardization
IT	information technology
ITPM	inspection, testing, and preventive maintenance
MOC	management of change
MSDS	material safety data sheets
OECD	Organization for Economic Cooperation and Development

OEM	original equipment manufacturer
ORR	operational readiness review
OSHA	Occupational Safety and Health Administration
P&ID	piping and instrumentation diagram
PHA	process hazard analysis
PSI	process safety information
PSKD	process safety knowledge and documentation
PSM	process safety management
PSSR	pre-startup safety review
PSV	pressure safety valve
PTFE	polytetrafluoroethylene
R&D	research and development
RAGAGEP	recognized and generally accepted good engineering practice
RBPS	Risk Based Process Safety
RCMS	Responsible Care Management System®
RFC	request for change
RIK	replacement-in-kind
RMP	risk management program

GLOSSARY

Authorization review. Approval mechanism for verifying that all identified hazards have been addressed and associated tasks have been performed prior to implementing a change.

Change. Any addition, process modification, or substitute item (e.g., person or thing) that is not a replacement-in-kind.

Change originator. Any individual who identifies the need for a change and initiates the MOC process through a request for change.

Classification review. Determination of which functions (e.g., engineering, safety) need to perform hazard reviews and authorization reviews for a change.

Closeout review. Approval mechanism for verifying that tasks required for a change have been completed. These tasks do not necessarily need to be performed prior to implementing the change.

Emergency change. A change needed in a situation where the time required for following the normal MOC procedure could result in an unacceptable safety hazard, a significant environmental or security incident, or an extreme economic loss.

Hazard review. Identification of (1) potential process safety problems (or other problems, such as environmental incidents, if the system scope includes them) to be resolved and (2) required controls to be implemented prior to and following a change.

Initial review. Preliminary determination of whether a proposed modification is worth pursuing and whether it is a change or a replacement-in-kind, based on MOC system definitions.

MOC coverage boundary. A physical, functional, or operational area of a facility or company that defines where an MOC system is implemented.

MOC coordinator. The individual responsible for the MOC system in all or part of a facility.

MOC documentation. Records that describe: the proposed change, the analyses performed to support the review and authorization of the RFC, any records of follow-up actions that were necessary to ensure that the change was completed as specified, and all other documents related to the RFC.

MOC performance. A determination using data that, when tracked, can help identify problems with MOC system operation and enhance continuous improvement efforts.

MOC system boundary. A management system activity at the "edge of inclusion" in the MOC system in which information, work products, or responsibility passes from the MOC system to the area of responsibility of an "adjacent" management system element (e.g., Operating Procedures element).

OSHA Process Safety Management, 29 CFR 1910.119 (OSHA PSM). A U.S. regulatory standard that requires use of a 14-element management system to help prevent or mitigate the effects of catastrophic releases of chemicals or energy from processes covered by the regulation.

Process safety information. Information pertaining to the properties of the hazardous chemicals used or produced by the process, the technology of the process, and the equipment in the process.

Process safety management. A management system that is focused on prevention of, preparedness for, mitigation of, response to, and restoration from catastrophic releases of chemicals or energy from a process associated with a facility or activity. "Process safety management" or "PSM", as used in these guidelines, is not meant to imply reference to the Occupational Safety and Health Administration's process safety management regulation (29 CFR 1910.119).

Replacement-in-kind (RIK). An item (equipment, chemicals, procedures, organizational structures, people, etc.) that meets the design specification, if one exists, of the item it is replacing. This can be an identical replacement or any other alternative specifically provided for in the design specification, as long as the alternative does not in any way adversely affect the function or safety of the item or associated items. For nonphysical changes (relating to procedures, personnel, organizational structures, etc.), no specification, per se, may exist. In these cases, the reviewer should consider the design and functional requirements of the existing item (even if nothing is written down) when deciding whether the proposed modification is an RIK or a change.

Request for change (RFC. A formal request to modify equipment, chemicals, procedures, organizational structures, staffing, and so forth. This can be done either using an RFC form or integrating RFC information into an existing work request/control document (e.g., maintenance work order).

Risk. A measure of potential loss (e.g., human injury, environmental impact, economic penalty) in terms of the magnitude of the loss and the likelihood that the loss will occur.

Risk analysis. The determination of a qualitative and/or quantitative estimate of risk based on engineering evaluation and mathematical techniques (quantitative only) for combining estimates of event consequences and frequencies.

Technical basis. An explanation of the proposed modification, including the reason(s) for performing the work, desired results, technical design, and appropriate implementation instructions.

Temporary change. A change that is implemented for a short, predetermined, finite period.

Written program. A written description of the roles, responsibilities, practices, procedures, and desired results associated with a management system for process safety. Most process safety management elements should have written programs to ensure consistent performance.

EXECUTIVE SUMMARY

Consistent and effective management of change (MOC) is one of the most important and difficult activities to implement in a company. MOC is important because uncontrolled changes can directly cause or lead to catastrophic events as well as degrade the quality of manufacturing operations. Formal MOC systems include administrative procedures for the review and approval of changes before they are made. This process helps ensure the continued safe and reliable operation of facilities.

The scope, level of detail, and complexity of an MOC system can have a significant impact on its success. MOC systems should be designed to fit the organizational structure, culture, and workforce of a facility. Well-designed systems are less likely to be used in a perfunctory fashion or circumvented. Having an inadequate system or one that is dormant is worse than having no system at all because facility management can be lulled into complacency, thinking that they are effectively managing change when change management is really not happening. MOC systems are also being adapted to deal with newly recognized, subtle sources change (e.g., organizational changes).

Several principles exist for successfully implementing MOC systems in a company or facility:

- ***Keep it simple, yet fit for duty.*** A modest system that works is better than an elegant one that does not.
- ***Obtain widespread acceptance and commitment.*** Solicit the opinions and concerns of all affected groups when developing a system.
- ***Field test the system prior to its official implementation.*** Debugging it early will pay off in the long run.
- ***Provide adequate training.*** Affected personnel should be educated on the existence of the system and their roles and responsibilities within it.

- ***Periodically monitor the effectiveness of the MOC system.*** Integrate the use of performance/efficiency metrics into real-time control of the system.
- ***Use audits and management reviews.*** Routinely monitor the MOC system to be sure the system is functioning as expected. A management system that is never reviewed will eventually degrade. Find ways to continuously improve your MOC procedures and practices.
- ***Demonstrate management leadership and commitment.*** Properly support the MOC program by providing adequate resources and making the hard decisions in favor of safety when MOC reviews indicate a problem. Like most aspects of process safety, MOC success begins at the top.

In general, an MOC system can address process safety issues and be applied to all operations involving the manufacture, use, or handling of hazardous substances or energy. However, the company should determine the physical areas of a facility where MOC is applied, the phases of a process life cycle for application (e.g., process development, design, construction, operation, decommissioning), and the sources of change (e.g., hardware, software, procedures, personnel, organizational). The level of detail used in an MOC system should be based on (1) the hazards or risk of the process, (2) the expected rate of use of the MOC process, and (3) the existing process safety culture at the location where the MOC system will be used.

These MOC Guidelines are not meant to represent the sole path for compliance with process safety regulations, nor is this book meant to establish new performance-based requirements for process safety. Nonetheless, in some sense, these MOC Guidelines do establish new risk-based expectations for process safety management and MOC.

The MOC element guidance is meant to be evaluated by companies that may elect to implement some aspects of these practices based on a thoughtful consideration of the risk-based design and implementation criteria. Not all companies even those with facilities in nearly similar circumstances may elect to adopt and implement the MOC activities in the same way. Company-specific and local circumstances may give rise to very different applications of MOC activities.

These guidelines can be used to establish new MOC systems or to improve existing systems. Please note that not all of the features described in these guidelines may be appropriate for all MOC systems.

1

INTRODUCTION

Management of change (MOC) is a process for evaluating and controlling modifications to facility design, operation, organization, or activities – *prior to implementation* – to make certain that no new hazards are introduced and that the risk of existing hazards to employees, the public, or the environment is not unknowingly increased.

MOC is one of the most important elements of a process safety management (PSM) system. Changes occur when modifications are made to the operation or when replacement equipment does not meet the design specification of the equipment it is replacing. Other, more subtle changes can occur when new chemical suppliers are selected, National Fire Protection Association hazard classifications change, procedures are modified, or site staffing and/or company organization is revised. Such changes, if not carefully controlled, can increase the risk of process operation and result in incidents.

MOC has been called the minute-by-minute risk assessment control system in plants and companies. The significance of MOC – or the lack of it – was never more apparent than in the Flixborough accident, as shown in Figure 1.1.[1] This watershed event involved a temporary modification to piping between cyclohexane oxidation reactors. In an effort to maintain production, a temporary bypass line was installed when the fifth of a series of six reactors was removed at a facility in Flixborough, England, in March of 1974. The bypass failed while the plant was being restarted after unrelated repairs on June 1, 1974, releasing about 60,000 pounds of hot process material, composed mostly of cyclohexane. The resulting vapor cloud exploded, yielding an energy release equivalent to about 15 tons of TNT. The explosion completely destroyed the plant, and damaged nearby homes and businesses, killing 28 employees, and injuring 89 employees and neighbors.

FIGURE 1.1 Flixborough Accident — Failure to Manage Change

No engineering support was available in the plant at the time of the accident. The temporary modification was constructed by people who did not know how to design large pipes equipped with bellows. As stated in the official report: "...they did not know that they did not know." An effective MOC system should have discovered the design flaw before the change was implemented, thus averting the disaster.

1.1 HISTORICAL PERSPECTIVE

Many companies have implemented MOC systems over the past 15 years. In 1989, the Center for Chemical Process Safety (CCPS) published its groundbreaking *Guidelines for Technical Management of Chemical Process Safety*, which included MOC as an element.[2] However, most of the initial chemical industry MOC implementation activity has been driven by two forces: (1) the Occupational Safety and Health Administration's (OSHA's) PSM standard and (2) quality initiatives.[3,4]

In 1993, the Chemical Manufacturers Association, now known as the American Chemistry Council (ACC), published the first comprehensive guidelines on MOC: *A Manager's Guide to Implementing and Improving Management of Change Systems.*[5] However, this treatise was not widely distributed. Since that time, many conference presentations have been given, journal papers written, and several additional texts completed on MOC; and yet the industry "thirst" for effective MOC practices remains.[6-7] More than ever before, companies recognize that insufficient control of changes plays a major role in accidents.

In addition, much has happened in the chemical industry since 1989 and a large amount of experience (good and bad) has been accumulated. Table 1.1 lists a number of events, happenings, trends, and experiences that CCPS considered as inputs to the development this book.

Given this industry experience, CCPS has developed these *MOC Guidelines* considering CCPS's new Risk Based Process Safety (RBPS) system approach (Chapter 2).[8] Table 1.2 lists the goals of these *MOC Guidelines* in serving identified industry needs.

As a result, companies can use these guidelines for any of the following activities:

- Implementing a company's first MOC system
- Diagnosing and correcting a defective MOC system
- Determining ways to continuously improve MOC effectiveness

1.2 MANAGEMENT OF CHANGE ELEMENT OVERVIEW

MOC reviews are performed at operating sites or in company corporate offices that are involved with capital project design and planning. MOC reviews focus on bona fide changes, not replacements-in-kind (RIKs). An employee first originates a change request. Then qualified personnel, normally independent of the MOC originator, review the request to identify any potentially adverse impacts. Based on this review, and after addressing any additional requirements, a responsible party either approves or rejects the change for execution. If the change is approved, it can be implemented. Before startup of the change, potentially affected personnel are either informed of the change or provided with more detailed training, if needed. Affected process safety information (PSI) is modified to reflect the change. Most of the time, these activities are completed prior to startup of the change.

TABLE 1.1. Things that Have Happened in MOC Since 1992

- More than 15 years of MOC experience, particularly with incidents for which failure of MOC was identified as a root cause
- Major increase in the use of electronic documentation of site information
- Emergence of MOC software applications
- Emergence of Web-based documentation sharing systems
- Company-wide MOC systems (involvement of non-local personnel in MOC reviews)
- Redistribution of PSM work to sites (lack of central monitoring of PSM/MOC)
- Downsizing and integration of MOC duties within production jobs
- Increased efforts to monitor MOC implementation via management reviews
- Organizational upheaval (divestitures, acquisitions lack of culture integration)
- Use of MOC in process areas not covered by regulatory standards
- Realization of the need for MOC for nontraditional types of changes
- PSM regulatory creep (broadening of the application for new change types and expanding the MOC work required)
- Expansion of the six-sigma approach and other productivity improvement initiatives, which has increased the workload associated with MOC systems involving subtle types of changes
- Accident investigations that have revealed the risk significance of previously under-considered sources of subtle change, such as organizational changes

TABLE 1.2. Goals of these *MOC Guidelines*

- Reduce the number of MOC related incidents and PSM audit findings
- Expand MOC into the process/project life cycle and nontraditional types of changes
- Tailor MOC systems to the facility size, perceived risk anticipated usage rate of the MOC system, and safety culture
- Monitor MOC performance at sites from afar, in real time, and cost effectively
- Quickly diagnose MOC problems without having to perform or wait for a PSM audit
- Make MOC systems more fault tolerant and resistant to circumvention or human error
- Monitor MOC performance and efficiency in a practical way
- Achieve better MOC results with fewer resources, if possible

The main product of an MOC system is a properly reviewed change request that is authorized, amended, or rejected. Ancillary products include modified PSI, change communication, and updated training records.

Companies and sites usually have written MOC procedures that apply to all work that is not judged to be an RIK. The results of the review process are typically documented on an MOC review form. Backup information provided to aid the review or generated by the review is usually kept for several years as a foundation for updates and process hazard analysis (PHA) revalidations. This information also provides an auditable record of the MOC implementation process.

1.3 MOTIVATIONS FOR MOC

Companies that manufacture, handle, store, or use hazardous chemicals are committed to effective MOC for a variety of reasons. In addition to a desire to promote employee and public safety and to protect the environment, motivations for MOC include the intent to comply with (1) ACC's Responsible Care® initiative, (2) government regulations requiring MOC systems, and (3) quality/environmental initiatives such as International Organization for Standardization (ISO) 9000/14000.[3, 5, 9-12]

PSM practices and formal management systems have been in place in many companies for more than 20 years. PSM is widely credited for perceived reductions in major accident risk and improved chemical industry performance. Nevertheless, many companies continue to be challenged by resource pressures, inadequate management systems (as evidenced by chronic deficiencies found in MOC audit results), and stagnant process safety incident performance, particularly involving MOC systems.

1.3.1 Internal Motivations

Inappropriate changes can affect employee and/or public safety, damage the environment, or result in significant business interruptions. They can also reduce product quality or increase production costs. The desire to decrease the occurrence of change-induced incidents and reduce the cost of doing business motivates companies to create effective MOC systems that will enable them to remain competitive, grow, and prosper.

Experience has demonstrated that inadvertent, unintended, erroneous, or poorly performed changes – changes whose risk is not properly understood – can result in catastrophic fires, explosions, or toxic releases. The 1974 explosion at Flixborough, England, described at the beginning of this chapter, was fundamental to the development of formal safety management systems, both in Europe and the United States. Table 1.3 gives examples of changes that could increase risk.

MOC systems call for implementation of formal administrative procedures that require reviews and approvals of proposed changes within designated areas of a site. The objective of MOC is to prevent changes in process chemistry and technology, equipment operations, maintenance, and supporting functions from introducing unacceptable risks. Inadequate reviews of proposed changes can result in the potential for certain changes to violate the design basis of carefully engineered systems or to increase the risk of processes that have operated safely for years.

1.3.2 Industry Initiatives

Several industry organizations have recommended the development of MOC procedures through various guidelines (Table 1.4).

TABLE 1.3. Examples of Changes that Should Be Managed or Could Increase Risk

Process equipment changes **such as materials of construction design parameters, and equipment configuration**
• Changing piping from carbon steel to stainless steel without considering the potential for pitting due to the presence of chlorides
• Replacing a reactor with one of equal volume but different length-to-diameter ratio without considering potential changes in vessel mixing and heat transfer characteristics
• Changing a vessel's service to a higher specific gravity material without considering the impact of the additional weight on the vessel support structure
• Changing a pump impeller to a larger diameter to increase capacity or head without considering the potential to (1) overpressure downstream equipment, (2) operate above PSV set pressures, or (3) cause pump cavitation because of suction side limitations
• Repairing a process leak via an engineered clamp without confirming that the pressure rating for the temporary repair is adequate for the service
• Replacing a metal wafered gasket with a Teflon gasket, which won't hold up to an external fire, on a temporary basis to make it through the weekend.
• Connecting the cooling system of a new reactor to an existing cooling tower, without assessing the impact of increased load on the tower
• Substituting plastic pipe for steel pipe without considering the potential for generating static electricity that could ignite flammable vapors or combustible dusts, or failure caused by lack of support, particularly at elevated temperatures
• Temporarily replacing a centrifugal pump with a positive displacement pump without considering the need for a reliable relief path in the downstream piping
Process control changes **such as instrumentation, controls, interlocks, and computerized systems, including logic solvers and software**
• Raising the trip point on a safety-related high level alarm beyond the safe operating limit established by prior safety analyses
• Permanently converting a 1-out-of-3 voted safety sensing system to a 1-out-of-2 system because one of the sensors has failed, which ignores the hardware fault tolerance of the safety system
• Replacing a transmitter that produces an analog output with one that produces a digital output without considering the failure modes associated with the new transmitter and the potential effect on the reliability of the associated interlock circuit
• Adding a new alarm within the DCS without considering the incremental impact for creating a process alarm overload situation for operators
Safety system changes **such as allowing process operation while certain safety systems are out of service**
• Adding an isolation valve beneath a pressure relief valve to make it easier to remove and test the relief valve without considering the management system required to be certain the valve is not inadvertently closed
• Replacing a building sprinkler system with a CO_2 system without considering the associated asphyxiation hazard
• Directing atmospheric relief valve discharges to an existing flare header without considering the impact on the flare header or the performance of other relief devices discharging into the header
• Replacing an explosion relief vent panel with a panel having a higher burst pressure to "prevent spurious openings"

TABLE 1.3. Examples of Changes that Should Be Managed or Could Increase Risk (cont'd)

***Site infrastructure changes,* such as fire protection, permanent and temporary buildings, roads, and service systems**

- Increasing the occupancy of the control room building without considering the increased risk of building occupancy
- Increasing the size of the chemicals warehouse without considering the impact requirements for sprinkler protection may have on the flow/pressure capability of the firewater supply
- Relocating a unit's control room to a remote location to reduce operator exposure to unit hazards, without considering the impact of decreased operator presence in the process area
- Temporarily closing a major site road because of interferences from a construction project or a maintenance turnaround without considering the impact on the accessibility of emergency response vehicles to certain portions of the facility
- Disbanding facility emergency response capabilities in lieu of support from municipal emergency response agencies without considering the response time and capabilities of such groups

***Operations and technology changes* such as process conditions, process flow paths, raw materials and product specifications, introduction of new chemicals on site, and changes in packaging**

- Increasing process throughput beyond the currently established unit nameplate capacity without considering the potential impact on relief system capacity requirements
- Temporarily bypassing a heat exchanger without considering low temperature embrittlement of downstream equipment
- Temporarily receiving a highly toxic material via tank truck instead of railcars without considering that more frequent connections and disconnections of unloading lines could increase the likelihood of process material releases
- Using a more reactive catalyst type than that recommended by the vendor without considering that the higher reaction rate may exceed the cooling capacity of the reactor, potentially leading to runaway reaction

***Changes in inspection, testing, and preventive maintenance, or repair requirements,* such as lengthening an inspection interval or changing the lubricant type used in a compressor**

- Postponing a unit turnaround beyond the design run time limit, resulting in exceeding the maximum allowable intervals for certain equipment tests and inspections
- Increasing maintenance intervals based on resource constraints without considering past operating experience
- Reassigning certain maintenance tasks from maintenance personnel to operators without providing the operators with appropriate procedures, tools, and training for their new responsibilities
- Changing the inspection method for unit piping thickness from ultrasonic to X-ray without considering the hazards associated with more frequent use of ionizing radiation in the unit

TABLE 1.3. Examples of Changes that Should Be Managed or Could Increase Risk (cont'd)

***Changes in procedures*, such as standard operating procedures, safe work practices, emergency procedures, administrative procedures, and maintenance and inspection procedures**
• Modifying operating procedures to reduce or eliminate operator rounds in an area without considering the benefits of operator presence, such as leak detection
• Changing previously established safety, quality, or operating limits in the operating procedure
• Moving from a hard-copy based operating procedure system to one where personnel access all procedures through the site intranet
• Abandoning the OEM manuals in lieu of site-generated maintenance procedures
***Organizational and staffing changes* such as reducing the number of operators on a shift, changing the maintenance contractor for the site, or changing from 5-day operation to 7-day operation**
• Relocating the site technical group to a remote central corporate location without considering the impact on their ability to provide support to the facility
• Changing from an 8-hour shift schedule to a 12-hour shift schedule without evaluating the potential effect of greater fatigue associated with longer shifts
• Replacing an operations unit manager without considering the training needs for the new unit manager
• Deciding not to replace a retiring corporate loss prevention expert who previously reviewed all relief system designs, or replacing the expert with an inexperienced engineer
• Realigning the corporate PSM auditing function, placing primary auditing responsibility at the site level, without considering the possible reduced expertise or independence of local auditors
***Policy changes*, such as changing the amount of overtime permitted**
• Liberalizing the limits on the amount of overtime that an individual can work each month without considering the possibility of worker fatigue, or reducing the amount of overtime without considering the impact on staffing emergency response teams
• Revising the facial hair policy to allow facial hair for some classes of employees who are perceived to have a reduced need to wear respiratory protection
• Adopting a new paperless document policy intended to manage all site documentation electronically, including review/authorization, access, and retention of PSM-related information on PHAs, procedures, MOCs, PSSRs, and training records
• Implementing a new corporate policy for selecting external equipment manufacturers/vendors and services that calls for a reverse auction and low-cost bidding process without consideration of the impact of non-standard equipment or less reliable equipment
• Changing the timing and means for shift change and turnover of operating control

TABLE 1.3. Examples of Changes that Should Be Managed or Could Increase Risk (cont'd)

***Other PSM system element changes*, such as modifying the MOC procedure to include a provision for emergency change requests**
• Reclassifying an area that currently requires a hot work permit as a designated area
• Revising the qualifications required for incident investigation leaders
• Eliminating a step in the approval of safe work permits that currently requires sign-off by the control room lead operator
• Modifying the way in which temporary trailer occupancy is controlled
***Other changes* including anything that "feels" like a change but does not fit in a change-type category that has been established for your facility; this "other type" should be in every MOC system**
• Adopting a new RAGAGEP on site, such as ISA 84.0104 standards for safety interlock life-cycle management
• Relocating a laboratory within an existing building
• Adding/deleting emergency response rolling stock (ambulances, etc.)
• Local municipalities/governments consolidating police, emergency medical service, and fire emergency response capabilities into one central location with enhanced communication and response technologies
• Changing the policy of using bicycles for onsite transportation

TABLE 1.4. Industry Initiatives to Implement MOC

- American Chemistry Council Responsible Care Management System® [9]
- American Institute of Chemical Engineers *Guidelines for Risk Based Process Safety* [8]
- American Petroleum Institute *Guidelines for Management of Process Hazards Recommended Practice 750* [13]
- Canadian Chemical Producers Association Responsible Care Program, *Manufacturing Code of Practices*
- GE Corporation, Six Sigma – The Road to Customer Impact

1.3.3 Regulatory Influences

Various U.S. and international government regulations require that changes to processes be reviewed. For example, the U.S. Congress has mandated that both OSHA and the Environmental Protection Agency (EPA) implement regulations that address accidents involving hazardous chemicals.[3,10] The regulations issued by both of these agencies include MOC requirements. In February 1992, OSHA adopted a regulation, *Process Safety Management of Highly Hazardous Chemicals* (29 CFR 1910.119), which requires MOC as a key element of a complete PSM program. Specifically, the OSHA PSM regulation [paragraph (l)] includes the following requirements:

- Develop written procedures for managing change
- Address the technical basis for each change
- Evaluate potential safety and health impacts for each change
- Define requirements for authorizing changes to be made
- Appropriately inform and train affected employees and contractors before changes occur

In addition, OSHA requires that MOC systems specify the appropriate time period for the change (e.g., a change that is permitted for only 1 week) and that PSI, procedures, and practices be updated, as necessary, when changes occur.

In June 1996, EPA finalized its risk management program (RMP) rule. The accident prevention program component of the RMP rule requires companies to develop MOC procedures.[10] These requirements are nearly identical to OSHA's MOC provisions, but they expand the evaluation to consider the potential offsite impacts of changes.

In addition to these federal regulations, various state process safety-related regulations specify MOC requirements. Companies should also consider these state regulations as they develop their corporate and local MOC programs.

Internationally, numerous legislations, regulations, and guidance documents require companies to address MOC (e.g., the EC Directive on Seveso, the UK COMAH regulations, OECD *Guiding Principles for Chemical Accident Prevention, Preparedness, and Response*).[14-16]

1.3.4 Quality Initiatives

ISO has established rigorous quality standards (i.e., the ISO 9000 series) that include MOC concepts for companies desiring to do business in the international marketplace. Specifically, ISO 9004, *Quality Management and Quality System Elements – Guidelines*, requires the documentation and authorization of all process changes. In addition, changes to work instructions, specifications, and drawings are to be controlled. Some purchasers of products have requested final approval of any MOCs related to that product to ensure that product quality is not compromised. ISO has also promulgated ISO 14000 on *Environmental Management Systems*, which also requires that changes be managed.

1.4 COMMITMENT REQUIRED FOR EFFECTIVE MOC SYSTEMS

Even though the concept and benefits of managing change are not new, the maturation of MOC programs within industry has been slow, and many companies still struggle with implementing effective MOC systems. This is partly due to the significant levels of resources and management commitment that are required to implement and improve such programs. MOC may represent the biggest challenge to culture change that a company faces. For example, seasoned engineers may feel as though an MOC process "second-guesses" their judgment, or operating managers may dislike having to "get permission" from others to make a change, even though they are the "experts."

Many companies have installed protocols for addressing changes without regulatory impetus because such controls represent sound business practices for achieving safety, quality, and environmental objectives. However, many of these protocols may not fully address the scope and depth that external guidelines and regulations now demand. That is, the MOC systems at many companies may lack the formal structure to help ensure that:

- Designs of site processes are well understood and documentation is up to date
- Proposed modifications are routinely evaluated for potential safety and health impacts before being implemented
- The level of detail for each review is appropriate for the potential hazard it poses
- The appropriate level of company management authorizes the changes
- Related activities required to safely implement the changes (e.g., training) are conducted
- Training of personnel on the changes is effective
- Records are maintained to document the changes

Developing an effective MOC system may require evolution in a company's culture; it also demands significant commitment from line management, departmental support organizations, and employees. Strong management commitment should include allocation of adequate resources for managing change and the willingness to modify existing management systems when necessary to accommodate MOC requirements. Only when management commitment is visibly demonstrated is it possible to obtain the widespread involvement and support essential to implementing an MOC system. In addition, to obtain the employee commitment necessary to make widespread employee involvement effective, management should provide effective orientation and training for all personnel (including contract personnel) involved in activities that can result from or be affected by changes.

1.5 ORGANIZATION AND USE OF THESE GUIDELINES

These *MOC Guidelines* are meant to be evaluated by companies who may elect to implement some aspects of these practices based on a thoughtful consideration of risk-based design and implementation criteria. Not all companies – even those with facilities in nearly similar circumstances – may elect to adopt and implement the MOC activities in the same way. Company-specific and local circumstances may give rise to very different applications of MOC activities based on the perceived needs, resource requirements, and existing safety culture of the facility.

These *MOC Guidelines* are not meant to represent the sole path for compliance with process safety regulations, nor is this book meant to establish new performance-based requirements for process safety. Nonetheless, in some sense, these *MOC Guidelines* do establish new risk-based expectations for PSM and MOC.

Companies can use the information provided in this book to help implement new MOC systems, repair defective systems, or improve mature systems using a life-cycle approach, including the following tasks:

- Design the MOC system
- Develop a written description of the system based on the design requirements
- Install the system
- Operate the MOC system over the life of the site
- Maintain the system and modify it as appropriate using information from audits and management reviews and through continuous improvement activities

This book devotes chapters and appendices (as appropriate) to each of these activities. Personnel creating a new MOC system or repairing/improving an existing one can consider the features described for each activity. Several appendices include additional information useful to those personnel.

Table 1.5 provides a list of perceived user needs and instructions on how to use this book to best meet those needs.

TABLE 1.5. Using ***Guidelines for Management of Change for Process Safety***

User Need Description	Sections to Review to Meet Needs
Want to know the basics	1, 2
Just getting started	1, 2, 3, 6, Appendices A, B and C
MOC system may be broken	1, 2, 3, 4, 5, 6, Appendices C, G, and H
Established system trying to get better	1, 2, 6, Appendices F and G
Understand MOC regulatory requirements	1, 4.5.4
Use MOC during process design	1, 2, 3, 4
Develop a corporate MOC policy	1, 2, 3
Develop an MOC awareness presentation	1, 2, 3, Appendix A
Improve audit protocol for MOC	1, 2, 3, 4, 5, Appendix E
Go from a paper system to an electronic MOC system	1, 2, 3, 4, Appendix D

Although managers and engineers can use these guidelines to implement, correct, and improve MOC systems at their sites, they can also be used by corporate personnel responsible for establishing company-wide standards or guidelines for MOC systems. In either case, the MOC implementation process described in this book allows company management to implement an MOC system that has a level of detail commensurate with the hazards associated with the facility and that is appropriate and workable for the site.

2

RELATIONSHIP TO RISK BASED PROCESS SAFETY

The Center for Chemical Process Safety (CCPS) has published *Guidelines for Risk Based Process Safety* (RBPS), a comprehensive look at the next-generation process safety management (PSM) system.[8] These management of change (MOC) guidelines are intended to be consistent with the principles in that book.

2.1 BASIC CONCEPTS AND DEFINITIONS

This chapter reviews terminology necessary for understanding how MOC systems fit within the RBPS management system and how readers can use these guidelines to help achieve accident prevention, preparedness, and response goals.

2.1.1 Process Safety and Risk

Process safety deals with the prevention of catastrophic releases of chemicals or energy from systems handling hazardous substances that could affect workers, the community, the environment, or business continuity. Risk deals with the lack of certainty about the ability to be accident-free and is best described by the following basic risk questions concerning a process or operation:

- What can go wrong?
- How likely is it?
- What are the impacts?

Based on the level of understanding of answers to these three basic risk questions and knowledge of regulatory and other constraints, a company can determine how it can best manage change in order to manage risk. Early in the life cycle of a process (i.e., conceptual design), limited information typically exists to answer all three of these questions – normally only enough information exists to understand the hazards of the chemicals/process. Once a process moves into the detailed design stage or is put into operation at a site, more detailed answers to these three questions can be discovered.

Understanding of risk helps a company decide how to shape its PSM activities. Even in a highly regulated environment, process safety professionals have a wide range of options to choose from when deciding how much technical rigor to incorporate into the PSM activities at their facilities. Sometimes this flexibility is limited by regulatory constraints, which define a minimum standard of performance for process safety activities. In some cases, an industry consensus standard or internal company requirement may shape or limit the process safety professional's design or improvement options. The range of options may be further constrained by corporate policies, standards, or guidelines.

Understanding risk is the most important part of a foundation for determining the type, capability, and dependability of the MOC system a facility needs.

2.1.2 Management Systems

Causes of chemical process accidents fall into one or more of the following categories:

- Technology failures
- Human failures
- Management system failures
- External circumstances/natural disasters

For many years, companies focused their accident prevention efforts on addressing technology and human factors. Incidents continued to occur despite industry efforts. In the mid-1980s, following a series of serious chemical accidents around the world, companies, industries, and governments began to focus on management systems (or lack thereof) as the underlying cause of these accidents. As a result, a large effort was launched to find ways to accelerate the industry adoption of a management systems approach to solving process safety problems.

Management system approaches had already begun to take root in the area of product quality, as evidenced by the establishment of various Total Quality Management frameworks. Moreover, the evolution of integrating manufacturing excellence into the business model has helped focus attention on boosting PSM performance.

A management system is a framework for getting work done in a dependable way over a long time. In the U.S., the introduction of these approaches prompted companies to initiate somewhat fragmented hazard analysis and equipment integrity efforts. Eventually, companies realized that an integrated management systems approach might be useful in focusing future accident prevention activities.

Management systems need to address certain issues in order to be comprehensive and dependable. Table 2.1 lists important issues that should be addressed in any management system. A PSM system that focuses on work activities to prevent, prepare for, mitigate, or respond to accidental releases should also address these issues – either in each individual PSM element [e.g., roles and responsibilities in an MOC or process hazard analysis (PHA written program] or in a single PSM element (e.g., auditing issues are all contained in the auditing element).

Whether designing or reconfiguring individual elements or the entire PSM system, the items in Table 2.1 should be used to ensure that the management systems issues that are essential for success are being addressed.

Because of the breadth and complexity of the activities within their scope, PSM systems are typically broken down into a layered hierarchy. The most basic level within a PSM system is the element. *MOC is an element within the CCPS RBPS system structure.* A written program for MOC should address all of the components in Table 2.1.

TABLE 2.1. Important Issues to Address in a Process Safety Management System
• Purpose and scope
• Personnel roles and responsibilities
• Tasks and procedures
• Necessary input information
• Anticipated results and work products
• Personnel qualifications and training
• Activity triggers, desired schedule, and deadlines
• Resources and tools needed
• Continuous improvement
• Management review
• Auditing

2.1.3 Life Cycles of Processes and Management Systems

Physical processes have life cycles consisting of several stages: conceptual design, research and development, detailed engineering design, procurement, construction, startup, normal operation, maintenance and turnarounds, and decommissioning. The names, numbers, and sequence of life-cycle stages vary across industries and companies; no commonly accepted set of descriptors exists. *MOC is an important activity in each life-cycle stage.* For simplicity, in this book CCPS chooses to use the following definitions for life-cycle stages:

- Process development
- Detailed design
- Construction and startup
- Operating lifetime
- Extended shutdowns
- Decommissioning

Like physical processes, management systems also experience life-cycle stages, even if a company does not explicitly recognize such stages. Thus, management systems should also be carefully designed, built, started up, operated, maintained, and eventually shut down or decommissioned.

2.1.4 Responses to Management System Problems

As PSM systems operate, they occasionally become defective, less effective, or fall into disuse. Facility management will typically diagnose and control the performance of its PSM system using a variety of means and sources of information. One typical approach is the use of an audit, whereby independent personnel evaluate the PSM activities to determine whether the PSM system is adequate and is being implemented in a dependable fashion. These audits can be resource-intensive and are typically performed at one- to three-year intervals. In between these audits, management is increasingly using metrics to monitor the PSM system on a more real-time basis.

PSM systems or elements that are found to be nonconforming (typically via PSM audits) – or even worse, chronically deficient – require correction. Companies that are fortunate enough to have PSM systems that run relatively problem-free still search for ways to improve their systems. *MOC is typically a very active management practice. Many companies focus a lot of attention on auditing and improving MOC systems.*

To help structure the discussion of PSM (or MOC) improvement, the following terms are defined below: performance, efficiency, effectiveness, and improvement.

Performance is reflected by the success with which the PSM/MOC work products from a specific PSM/MOC activity meet the company-defined standard for quality, thoroughness, and timeliness. PSM/MOC performance can be measured by outcome-oriented event indicators (e.g., incident rates) or process-related leading indicators (e.g., rate of improperly performed MOC activities).

On a company level, event indicators may be sufficient to provide an idea of where the company is going with respect to process safety; however, their power to discriminate and diagnose is limited. But on a site or process level, these statistics are not enough to help a company determine how close to the edge it is and where improvements need to be made. On the local level, PSM/MOC element leading indicators are one of the few ways that show promise in helping companies monitor the risk-health of their facilities.

Efficiency is reflected by the amount of resources used to create the desired PSM work product. Typically, resources are expressed in monetary terms or in terms of time spent in creating the work product. An adequate work product that costs less to make than it did last year is said to have been created more efficiently.

Effectiveness, therefore, is defined as the functional combination of performance and efficiency:

Effectiveness = function of [Performance & Efficiency]

To improve PSM/MOC effectiveness, a company can attempt one or more of the following:

- Achieve better results with no increase in costs
- Reduce costs while maintaining the same level of performance
- Improve performance and increase efficiency at the same time

Improvement efforts can address performance issues, efficiency issues, or both. Continuous improvement implies that the improvement activity is accomplished on a more regular, rather than episodic, basis. Thus, continuous improvement in PSM/MOC effectiveness must embody (1) regular, consistent activities and (2) tangible, positive changes in performance, efficiency, or both.

The following sections describe the RBPS system and the MOC system hierarchy.

2.2 OVERVIEW OF THE RBPS SYSTEM

An RBPS system addresses four accident operation pillars: (1) committing to process safety, (2) understanding hazards and evaluating risk, (3) managing risk, and (4) learning from experience. To manage risk, facilities focus on three aspects:

- Disciplined operation and maintenance of processes that pose residual risk and their associated protective systems
- Controlling changes to those processes and protective systems to avoid inadvertent risk increases
- Preparing for, responding to, and managing incidents that do occur

Efforts to control change-induced risk revolve around two RBPS elements: management of change and operational readiness. This section covers the attributes of an effective MOC system.

2.2.1 Risk Based Process Safety Management System Approach

RBPS is founded on the principle that appropriate levels of detail and rigor in process safety practices should be premised on the following three factors:

- Current understanding of the risk of the processes on which the process safety practices are focused
- Level of demand for the process safety activity (e.g., the number of changes that need review per month) and the sustainable resources available to support implementation over the life of the facility
- Existing company culture within which the process safety practices will be implemented

In this risk-based, layered approach, the right level of practices can be designed and implemented in a way that (1) optimizes PSM performance, efficiency, and effectiveness and (2) avoids gaps, inconsistencies, overwork, underwork, and associated process safety risks and economic risks.

Process safety professionals may have a wide range of options to choose from when deciding how much technical rigor to incorporate into their company/facility PSM activities. Sometimes this flexibility is limited by regulatory constraints, which define a minimum standard for pursuit of the process safety activity. In some cases, an industry consensus standard or internal company requirement may shape or limit a company's MOC system design/improvement options.

In either case, these requirements may be written in a prescriptive form or in a performance-based fashion. Prescriptive requirements state precisely how the process safety activity is to be conducted and what the activity is to produce. Performance-based requirements are more flexible because they specify only what is to be accomplished and leave the method for generating the desired results up to the company/facility or the process safety professional in charge of the activity.

A main focus of the RBPS approach is to help process safety professionals build and operate more effective PSM systems by providing guidance on how to design or improve a specific process safety activity so that the energy put into the activity is sufficient to meet the anticipated needs for that activity. In this way, limited company resources can be focused elsewhere to generate improved safety and economic performance.

Higher-risk situations usually require a more formal and thorough implementation of an MOC protocol (e.g., a detailed written program that specifies exactly how changes are identified, reviewed, and managed). Companies having lower-risk situations may appropriately decide to manage changes in a less rigorous fashion (e.g., a general policy about managing changes implemented by trained key employees using informal practices).

Facilities that experience high demand for managing changes may need greater specificity in the MOC procedure and greater allocation of personnel resources to fulfill the defined roles and responsibilities. Lower-demand situations allow facilities to operate an MOC protocol with greater flexibility.

Facilities with sound safety cultures generally have MOC procedures that are more performance based, allowing trained employees to use good judgment when managing changes in an agile system. Facilities with an evolving or uncertain safety culture generally require more prescriptive MOC procedures, more frequent training, and stronger command and control management system features to ensure disciplined MOC implementation.

2.2.2 Risk Based Process Safety Elements

Table 2.2 lists the elements in the CCPS RBPS model.[8]

2.2.3 RBPS System Design Hierarchy

The level of rigor that any particular company or facility applies to establishing or improving an MOC system should be based on the RBPS criteria: perceived hazard/risk, demand for resources, and culture. The following sections provide an overview of MOC practices that are in use in industry today. Increasingly greater detail is provided as one goes deeper into the MOC element structure given in the RBPS guidelines book (summarized in Appendix B of this guideline), which is organized as follows:

- ***Element*** (e.g., management of change)
- ***Key Principle*** (e.g., identify potential change situations)
- ***Essential Feature*** (e.g., all sources of change are managed)
- ***Possible Work Activity*** (e.g., develop a list of areas to which MOC applies)
- ***Implementation Options*** (e.g., an MOC coverage list is maintained and communicated)

TABLE 2.2. CCPS Risk Based Process Safety Elements

Commit to Process Safety
1. Process Safety Culture
2. Compliance with Standards
3. Process Safety Competency
4. Workforce Involvement
5. Stakeholder Outreach
Understand Hazards and Evaluate Risk
6. Process Knowledge Management
7. Hazard Identification and Risk Analysis
Manage Risk
8. Operating Procedures
9. Training and Performance
10. Safe Work Practices
11. Asset Integrity and Reliability
12. Contractor Management
13. Management of Change
14. Operational Readiness
15. Conduct of Operations
16. Emergency Management
Learn from Experience
17. Incident Investigation
18. Measurement and Metrics
19. Auditing
20. Management Review and Continuous Improvement

The following section discusses only the MOC key principles and essential features. Additional details about possible work activities are provided in Chapter 15 of the *RBPS Guidelines* and in Appendix B of this book.

2.2.4 Key Principles and Essential Features of MOC Systems

A company should address the following MOC key principles:

- Maintain a dependable MOC practice
- Identify potential change situations
- Evaluate possible impacts
- Decide whether to allow the change
- Complete follow-up activities

Section 2.1 of this guideline defines the generic requirements of a management system (roles and responsibilities, scope, task procedures, etc.).

Readers should keep these requirements in mind as they seek to implement a comprehensive MOC system in a risk-appropriate fashion. Some facilities may decide to implement an MOC system at the key principle level of rigor. Other facilities may decide that greater rigor is required, and they explicitly implement the essential features for each key principle by identifying effective work activities to accomplish each essential feature in the MOC system. Following is a brief description of each of the MOC key principles and a list of the essential features that support each key principle.

Maintain a Dependable MOC Practice

If a PSM activity is important enough to have been identified as something that should be done, then it is likely that the company/facility will want the activity to be performed in a fashion that is consistent over the life of the facility. In order for an MOC practice that applies to a variety of people and situations to be executed dependably throughout a facility, the following essential features should be considered:

- Establish consistent implementation
- Involve competent personnel
- Keep MOC practices effective

Identify Potential Change Situations

Modifications cannot be evaluated unless they are known. Companies/ facilities should implement effective means of identifying the types of modifications that are anticipated and the sources/initiators of these modifications. In order for an MOC system to address all potentially significant change situations, the following essential features should be considered:

- Define the scope of the MOC system
- Manage all sources of change

Evaluate Possible Impacts

Once potential change situations are identified, they can be evaluated using an appropriate level of scrutiny to determine whether the change introduces a new hazard or exacerbates the risk of an existing one. In order for companies/ facilities to adopt and implement appropriate review protocols for relevant change types, the following essential features should be considered:

- Provide appropriate input information to manage changes
- Apply appropriate technical rigor for the MOC review process
- Ensure that MOC reviewers have the appropriate expertise and tools

Decide Whether to Allow the Change

Once a change has been reviewed and the hazard/risk evaluated, management can decide whether to (1) approve the change for implementation as requested and thus accept any associated risk, (2) require amendment to the change request or the implementation process, (3) require that a more rigorous hazard evaluation be conducted, or (4) deny the change request. In order for companies/facilities to adopt and implement appropriate MOC approval protocols, the following essential features should be considered:

- Authorize changes
- Ensure that change authorizers address important issues

Complete Follow-up Activities

Once a change is authorized, it is released for implementation. Typically, the execution of a change is performed via work practices under other RBPS elements (e.g., mechanical integrity, operating procedures, safe work practices) by facility personnel or contractors involved in design, engineering, construction, operation, or maintenance. Prior to startup of the change (i.e., exposure of personnel to the modified situation, which could create new hazards or increase risk), certain activities may be required by the MOC procedure or the reviewers/authorizers (e.g., update process drawings, train affected personnel, implement required risk control measures).

Sometimes action items may be deferred until after startup; these items should be minimized and carefully tracked to completion to avoid potential failure to implement them. In order for companies/facilities to ensure that approved changes are properly followed up on, the following essential features should be considered:

- Update records
- Communicate changes to personnel
- Enact risk control measures
- Maintain MOC records

Chapters 3 and 4 of this book provide insights into how to design and develop an MOC system containing work activities to address each of the key principles and essential features mentioned above. Chapter 5 addresses how to diagnose and correct a seriously defective MOC system. Chapter 6 addresses how to improve the effectiveness of an existing, mature MOC system.

Note: The possible work activities, implementation options, and effectiveness improvement ideas found in the RBPS guidelines book and in the MOC system design tool described in Appendix B of this book may not be appropriate for every situation. Management should evaluate its own circumstances and determine the extent to which these activities are appropriate.

2.2.5 Interaction among MOC and Other RBPS Elements

The MOC system interacts with many other PSM elements because it is the day-to-day risk "watchdog." Many elements provide inputs to the MOC system, and the MOC system provides work products or action item requirements that will be executed by other RBPS elements as a result of authorized change requests. Table 2.3 lists the interactions that the MOC system typically has with other RBPS elements.

In addition, the MOC element may interact with other non-PSM management systems. For example, some companies may use their PSM MOC system as a way to manage changes unrelated to process safety issues (e.g., security, environmental, quality). In addition, depending upon the life-cycle stage at which changes are managed, the MOC system may interact with other systems or activities, such as project management, budgeting, and product development.

TABLE 2.3. MOC Inputs and Outputs

RBPS Element	Inputs to MOC from the Element	Outputs from MOC to the Element
Process Knowledge Management	• Chemical/process hazard information • Drawings • Equipment specifications • Safe operating limits • Safety system definitions	• Updates to all relevant process safety information, knowledge, and records
Hazard Identification and Risk Analysis	• Indication of process/activity risk • Risk tolerance criteria • Safety systems • Recommendations needing to be managed as changes	• Results of MOC hazard evaluation
Training and Performance	• Job qualifications • Staffing (number, composition, and required competencies)	• Information on changes to inform or train potentially affected contractor personnel • Changes to all process safety knowledge and documentation
Operating Procedures	• Operating procedures	• Changes needed to affected operating procedures
Asset Integrity and Reliability	• Maintenance procedures • ITPM frequencies • Personnel qualifications	• Updates to affected maintenance procedures, frequencies, and personnel
Safe Work Practices	• Safe work practice procedures • Criteria for applying procedures	• Updates needed to affected procedures, application criteria, and personnel
Operational Readiness	• Items discovered during a PSSR that require change to the process prior to start-up	• Change situations requiring PSSR • Results of MOC hazard evaluation • Risk control measures mandated by MOC review process
Contractor Management	• Qualification requirements • Training requirements	• Information on change to inform or train potentially affected contractor personnel • Changes to all process safety knowledge and documentation • Change implementation timing

3

DESIGNING AN MOC SYSTEM

When establishing objectives for its management of change (MOC) system, a company should consider applicable regulatory requirements and local facility needs. These objectives can be organized into a *design specification*, focusing subsequent development efforts for the MOC system and helping ensure that the system meets management's expectations. A formal design specification for a management system such as MOC may not always be needed to communicate management expectations, but some method of recognizing and addressing management's desires should be considered.

The MOC system design specification should address the following features:

- Terminology
- Implementation context
- Roles and responsibilities
- Scope of the system
- Interfaces with other company practices and programs
- Requirements for review and authorization
- Guidelines for key MOC issues
- Guidelines for making the MOC system easy to monitor

Chapter 2 discussed the Risk Based Process Safety (RBPS) element framework involving key principles, essential features, possible work activities, and implementation options. Appendix B provides a framework that sets forth in expandable fashion the various layers of detail/rigor that one could incorporate into an MOC system design. This tool should be used with the material in this chapter and in Chapter 4 to design and develop an MOC procedure. The following sections discuss each of the MOC system design features listed above.

3.1 ESTABLISHING TERMINOLOGY

A company needs to establish appropriate and consistent terminology to help minimize confusion during implementation and operation of an MOC system. This section defines the terminology used throughout this book to provide a common language for the reader. (A more complete list of suggested MOC terminology is provided in the Glossary.) While adopting the terminology presented here may be appropriate for some companies, it is more important to for company management to ensure that the definitions used in the design specification and other MOC system documents are consistent with terminology used in process safety management (PSM or related management systems).

The following terms are used in this book:

Replacement-in-kind (RIK). An item (equipment, chemicals, procedures, organizational structures, people, etc.) that meets the design specification, if one exists, for the item it is replacing. This can be an identical replacement or any other alternative specifically provided for in the design specification, as long as the alternative does not in any way adversely affect the function or safety of the item or associated items. For nonphysical changes (relating to procedures, personnel, organizational structures, etc.), no specification, per se, may exist. In these cases, the reviewer should consider the design and functional requirements of the existing item (even if nothing is written down) when deciding whether the proposed modification is an RIK or a change.

Change. Any addition, process modification, or substitute item (e.g., person or thing) that is not an RIK.

Request for change (RFC). A formal request to modify equipment, chemicals, procedures, organizational structures, staffing, and so forth. This can be done either using an RFC form or integrating RFC information into an existing work request/control document (e.g., maintenance work order).

Technical basis for change. An explanation of the proposed modification, including the reason(s) for performing the work, desired results, technical design, and appropriate implementation instructions. Often included on the RFC form, the technical basis for change should be of sufficient detail to allow appropriate supervisory, technical, and management review, including addressing the following questions:

- What is to be changed and how?
- What will be achieved by the change?
- How will the change achieve the intended goal?
- Is the change safe to make and why?

Emergency Change. A change that must be performed in a true emergency because of any of the following situations:

- The process must be changed to correct a deficiency that would cause a hazardous condition (i.e., an immediate threat to the safety of site personnel or the public)
- The process must be changed to prevent an immediate environmental release
- The process or facility must be changed to address impending external threats that could result in a release, such as natural disasters (e.g., hurricanes, floods), extreme temperatures (e.g., unusually cold weather in a warm climate), or imminent security risks
- The process must be changed to prevent an extreme economic loss (e.g., product loss or spoilage, catalyst degradation, business interruption, loss of market share)
- The process would be in jeopardy of severe financial loss from not providing product to customers because of equipment failure or unforeseen design errors

Caution: Companies choosing to define an emergency using an economic driver should closely monitor MOC system implementation to ensure that employees do not abuse the use of the normally less intensive and time-consuming emergency change provisions for the sake of convenience.

Temporary change. A change that is intended to exist for a short, predetermined, finite period. Temporary MOC procedures tend to follow the same work process as permanent changes, but they should be used only as long as the situation warrants since temporary changes may incur a higher level of short-term risk. After a short-term implementation period (e.g. 90 days), one of the following must occur: (1) a new permanent MOC must be initiated for review using data from the temporary change as justification, (2) the system must be returned to its original condition, or (3) the temporary change can be extended, with or without further review. Normal practice is to put limits on the number and/or duration of administrative extensions of temporary changes. Extensions or renewals of temporary changes without further review should be carefully considered and avoided if possible.

MOC documentation. Records that describe: the proposed change, the analyses performed to support the review and authorization of the RFC, any records of follow-up actions that were necessary to ensure that the change was completed as specified, and all other documents related to the RFC.

MOC terminology should be consistent throughout a facility to avoid possible miscommunication to site personnel of their responsibilities and management's expectations. Companies may need to develop additional terminology for use in their site-specific MOC programs.

3.2 DETERMINING THE IMPLEMENTATION CONTEXT

The design of an MOC system should consider the situation in which the system is intended to operate. The situation can be characterized generally by the life-cycle stage of the facility and the anticipated MOC system use rate.

3.2.1 Life-cycle Application

In this book, CCPS uses the following process life-cycle stage terms:

- Process development
- Detailed design
- Construction and startup
- Operating lifetime
- Extended shutdowns
- Decommissioning

MOC systems at early life-cycle stages (e.g., process development) are typically simpler and less structured than those associated with mature facilities (e.g., startup, operating life). At later, mature stages, facilities typically have much more process information upon which to base change review decisions, and the risk associated with change is more direct and tangible; consequently, MOC systems are comparatively more complex. MOC systems at end-of-life stages (e.g., decommissioning) tend to revert back to more simple procedures because of the uncertain nature of the work needed to permanently shut down the facility.

Like physical processes, management systems also experience life-cycle stages, even if a company does not explicitly recognize such stages. Thus, management systems are designed, built, started up, operated, maintained, and eventually may be shut down or decommissioned. Such management system implementation concepts should be considered when establishing an appropriate MOC system for the applicable life-cycle stages. Such maintenance or decommissioning of an MOC system should also be evaluated using an appropriate review process.

3.2.2 Considerations for MOC Systems in Non-traditional Activities

The need to manage change is not limited to operating plants. New hazards can be introduced or known risks can be unintentionally increased during every phase of a process life cycle, at locations that are not an operating site, or in non-traditional activities, such as the following:

- Research and development laboratories
- Process development centers
- Engineering design offices, including those of contractors
- Equipment fabrication yards
- Long-term in situ shutdown/mothballing of equipment
- Demolition
- Equipment preservation and storage

At any of these locations, or during any of these times or activities, a typical plant MOC procedure may not be applicable, appropriate, or appreciated.

What makes the need for managing change in these situations and, therefore, the design of the MOC procedure, different? The following are some aspects that differentiate early or late life-cycle circumstances from a normal plant situation:

- The number or frequency of changes may be vastly different
- The available information upon which to base an MOC hazard review may be much less or very different
- The types of changes may be different – and are likely to be more subtle
- The disciplines necessary to review a change are different
- The people available to approve a change are different
- The time frame for reviewing or implementing a change may be much different
- The tools or techniques needed to properly evaluate a change may be different and will be a strong function of the available information
- Access to information may be different (e.g., old paper records versus electronic documentation)
- Different companies may be involved
- Follow-up needs may be different

Although the work processes may be similar, the people, information, and techniques used in each basic MOC review step will likely be different at each life-cycle stage. For example, early life-cycle MOC work processes often use non-plant personnel, are based on more qualitative information, use less exhaustive hazard evaluation methods, and are carried out by fewer people.

Later life-cycle MOC procedures (e.g., demolition) will have some of those same characteristics, although the procedures will likely be carried out by operating site personnel. Table 3.1 outlines some considerations for designing a non-traditional MOC procedure.

3.2.3 Establishing MOC System Design Parameters

The need for change does not occur at regular intervals; change requests are random, or at least episodic. More changes can be expected in the detailed design to near-end-of-life stages of facilities. Fewer, less complex changes typically occur in early and late life stages. In each case, the MOC system designer should anticipate how the MOC system will be used during the applicable life-cycle stage to ensure that it is fit for duty considering the desired technical rigor and efficiency. Table 3.2 lists some considerations that should be addressed when designing an MOC system.

The design complexity of an MOC system should consider some or all of the parameters listed above. A management system is no different from a physical process system in that, if it is stressed beyond its design limits, it is prone to failure. Considering these factors during design at each life-cycle stage will help ensure a high-performing and efficient MOC system throughout the life of the facility.

TABLE 3.1. Considerations for Designing an Early or Late Life-cycle MOC System

MOC Resource Aspect		
Available Information	**Disciplines/People**	**Techniques/ Methods**
• More qualitative • Less equipment specific • Fewer change categories • Limited information available early in life • Less structured work-generation information (no work order system) • Fewer records to update very early in life and very late in life • Limited or no training necessary	• Chemists and designers early in life • Construction and maintenance engineers later in life • Fewer people involved in reviews • Parallel reviews likely	• Review procedure less detailed • Temporary changes unlikely • Emergency changes possible in later-life MOC systems • Less rigorous methods, such as hazard checklist or what-if analysis • Multiple sign-offs likely in early life • Single sign-offs likely later in life • Fewer closeout tasks likely later in life • Generates fewer records updates later in life

TABLE 3.2. MOC System Design Considerations

MOC System Issue	Description
System capacity	Number of MOC reviews (total or by type) that can be handled at the same time
MOC rate	Number of MOC reviews (total or by type) that are conducted on a daily, weekly, or monthly basis
Total reviews	Number of MOC reviews conducted over a long period
Completion/residence time	Actual or average amount of calendar time required from origination to completion/closeout of an MOC review
Anticipated backlog	Number or average age of MOC reviews that are late or not expected to be completed by the intended or desired change implementation date
Surge capacity	Increase in MOC rate or system capacity that can be sustained for short durations (e.g., a 2-week turnaround), typically using increased resources
Level of approval for MOC	Number and level of MOC approvers on and off site (area level, site level, off site [business or corporate level])
Available resources	Number of people, disciplines, job functions, or man-hours available for participating in the MOC review process
Level of computer literacy	Computer literacy of affected personnel (their ability to access MOC information or to sign off on MOC reviews electronically may determine the effectiveness of a paper system versus an electronic system)
Emergency change request needs	Anticipated rate of need for emergency change requests

3.2.4 RBPS Design Criteria

Chapter 2 described the RBPS strategic approach to designing PSM systems. The three design criteria below should be considered, along with the life-cycle stage and design parameters, when developing an appropriate MOC procedure that is fit for its intended use. The following items describe the influence of RBPS criteria on MOC system design:

- *Perceived risk.* Higher-risk situations usually require more formal and thorough implementation of an MOC protocol (e.g., a detailed written program that specifies exactly how changes are identified, reviewed, and managed). Companies having lower-risk situations may appropriately decide to manage changes in a less rigorous fashion (e.g., a general policy for managing changes that is implemented via informal practices by trained key employees).
- *Demand for resources.* Facilities that experience high demand for managing changes may need greater specificity in the MOC procedure and greater allocation of resources to fulfill the defined roles and responsibilities. Lower-demand situations allow facilities to operate an MOC system with greater flexibility.

- *Culture.* Facilities with a sound safety culture might choose to implement MOC procedures that are more performance based, allowing trained employees to use good judgment when managing changes in an agile manner. Facilities with an evolving or uncertain safety culture may require more prescriptive MOC procedures, more frequent training, and stronger command and control management system features to ensure disciplined MOC implementation.

Carefully considering the life-cycle stage, RBPS criteria, and MOC system design parameters will help ensure that the MOC system is as fit for the purpose as possible.

3.3 DEFINING ROLES AND RESPONSIBILITIES

Implementing an effective MOC system in a facility requires action by many different departments and individuals. The specific assignments of responsibility and authority may be different from location to location. For example, at a large facility, more than one person may be assigned full-time responsibility for some of the functions listed below. On the other hand, a small facility may have a single individual who performs many of the functions described below. Also, the MOC roles might not be full-time jobs, even for the MOC coordinator, unless the facility experiences a large number of changes.

The design specification should describe the titles and roles for key personnel in the MOC system. The following are generic roles and responsibilities associated with implementation of MOC systems:

Senior management. Senior managers at a site establish basic criteria for reviewing changes at the site. These managers, with input from the PSM manager, establish the specifications for the MOC system. The managers' most important decision is often the level of authority that will be necessary for approving each type of change. They may also specify the scope of the MOC system (e.g., they may choose to implement MOC more widely than is specified by regulatory requirements alone).

Process Safety Management manager. The PSM manager has responsibility for guiding the overall development of PSM element systems at the site and ensuring that these systems meet applicable requirements [e.g., the Occupational Safety and Health Administration's (OSHA's) PSM regulation, American Chemistry Council's (ACC's) Process Safety Code]. In addition, the PSM manager works to meld the individual PSM element systems (including MOC) into a cohesive PSM program. For example, at some facilities the PSM manager closely monitors the coordination of MOC and pre-startup safety review (PSSR) procedures.

MOC coordinator. This individual directs the activities associated with the MOC system and is often responsible for leading the development, installation, operation, and maintenance of the MOC system, including the MOC system procedures and records. The MOC coordinator also (1) helps define review procedures for changes that do not clearly fit into prescribed MOC categories, (2) serves as coach, counselor, and trainer to those implementing the MOC system, and (3) is often the final authority for deciding whether proposed work is a change or an RIK. Because of the importance of this position, companies should consider assigning qualified substitutes who are able to carry out these duties should the MOC coordinator be absent or unavailable.

MOC system development team. Under the direction of the MOC coordinator, this temporary group creates the MOC system procedures based on the MOC system design specification provided by management. Ideally, these individuals should be selected from a cross-section of company/facility departments (operations, maintenance, engineering, technical services, safety, etc.). One key to establishing a successful team is to enlist people (1) from several different organizations, (2) with different types and levels of experience, and (3) with specific day-to-day involvement in identifying, approving, and making changes (e.g., operators, maintenance planners and technicians, frontline supervisors, process engineers). For a small site, a single individual might conduct the development; however, other personnel representing a cross-section of perspectives and experience should then review the draft procedure. The development team for an MOC system will likely be different for early life-cycle stages and later life-cycle stages.

Change originators. These individuals (e.g., operators, maintenance technicians, frontline supervisors, inspectors, process engineers) typically identify needs and initiate requests for changes. Originators should propose only those changes that they believe can be implemented with manageable safety and health impacts. The originator's description of a proposed change should provide enough detail to allow for adequate evaluation during the MOC process. In many cases, the originator may be responsible for (1) developing, (2) assigning responsibility for developing, and/or (3) stewarding the development of the MOC package, which includes ensuring that all necessary supporting technical studies, design calculations, drawings, and specifications are completed and provided along with the RFC. The originator should classify the change for MOC review (including any special circumstances, such as temporary changes or emergency situations) and submit the RFC to a designated initial reviewer for that type of change request (e.g., process area supervisor for operational changes).

Designated initial reviewer. The designated initial reviewer determines whether (1) a change is truly needed and feasible and (2) the originator's classification of the modification is appropriate. These determinations often involve consultation with technical experts and other individuals. The initial reviewer is most often the supervisor of the person requesting the change. Many facilities assign qualified substitutes to carry out these duties should a primary reviewer be absent or unavailable. The initial reviewer is often responsible for ensuring that the proposed change does not violate appropriate financial or administrative protocols. The initial reviewer may also determine the level of additional review that is required.

RFC reviewers. The RFC reviewers must analyze a potential change for hazards before the change is implemented. Different types of reviewers may be required, depending upon the category and risk significance of the change involved. For example, a purchasing representative does not need to review a requested change in operating parameters for a unit. However, the purchasing representative may need to review a requested change in the quality control requirements for purchased material. In addition, the purchasing department may need to initiate the RFC if RIK is not achievable. These reviewers may work alone or as a team, and they may use formal hazard evaluation techniques to aid their assessment of potential safety impacts. Industry guidelines are available that discuss the use of hazard evaluation techniques for various purposes, including review of changes.[17-18] The type and rigor of the review may be risk based.

RFC authorizers. These individuals consider the results of each RFC review and (1) approve the change for implementation as requested and thus accept any associated risk, (2) require amendment to the change request or the implementation process, (3) require that a more rigorous hazard evaluation be conducted, or (4) deny the change request. Small sites and situations involving simple types of changes may have only a single, experienced individual designated to authorize changes.

Situations involving complex changes or high hazard levels may require that more than one person approve the change for implementation. If company management determines that more than one RFC authorizer is required for a particular type of change situation, the authorizers are usually chosen from different departments (e.g., operations, engineering, maintenance) in order to provide a multidisciplinary review and to help ensure the review's independence. In some situations, the RFC reviewers and the RFC authorizers may be the same people.

All employees. The effectiveness of all MOC systems ultimately depends upon the employees' commitment to identifying potential change situations and following the appropriate change review procedures. *Because of the need for employee commitment, educating all affected site employees about (1) the goals of the MOC program, (2) what constitutes a change, (3) their individual responsibilities under the MOC program, and (4) the identity, responsibility, availability, and authority of each MOC system participant is vitally important.*

3.4 DEFINING THE SCOPE OF THE MOC SYSTEM

The design specification created by facility management should define the scope of the MOC system, including the:

- Physical facility areas for which the MOC review protocols will be implemented
- Types of changes that will be evaluated using the MOC system
- Boundaries and intentional overlaps with other elements or administrative systems

3.4.1 Physical Areas for which MOC Will Be Implemented

To help ensure consistent application of MOC requirements among sites, corporate process safety managers may want to provide initial guidance about scoping considerations for use by individual sites. Also, in establishing the MOC program scope, site management may want to consult with corporate personnel or other company sites to determine which, if any, regulatory or other MOC obligations apply to their sites or processes. In determining which process areas require MOC, site management should also remember that local regulatory requirements might affect the definition of these areas.

Some companies apply MOC fenceline-to-fenceline to standardize on a single set of requirements site-wide.

3.4.2 Types of Changes to Be Managed

When defining the scope of a site MOC system, companies should consider including the types of changes listed in Table 3.3.

The MOC system should address any changes (including additions and deletions) to a process or its supporting systems. However, MOC review protocols established by a site *do not* apply to those actions that are deemed to be RIKs. Appendix A presents some examples of changes and RIKs for various classes of modifications. The actual MOC review protocol may be different for various categories of change and may use different RFC forms, as long as the same goals outlined for the MOC program are achieved. However, for consistency and efficiency, having the fewest different protocols and forms possible, while still meeting the need for thoroughness in reviewing anticipated change types, is best. Appendix C provides some examples of MOC review processes.

TABLE 3.3. Examples of Changes that Should Be Considered for Inclusion in the Scope of an MOC System Design

- ***Process equipment changes***, such as materials of construction, design parameters, and equipment configuration
- ***Process control changes***, such as instrumentation, controls, interlocks and computerized systems (including logic solvers and software)
- ***Operations and technology changes***, such as process conditions or limits, process flow paths, raw materials and product specifications, introduction of new chemicals on site, and changes in packaging
- ***Changes in procedures***, such as standard operating procedures, safe work practices, emergency procedures, administrative procedures, and maintenance and inspection/test procedures
- ***Safety system changes***, such as allowing process operation while certain safety systems are out of service
- ***Changes in inspection, testing, preventive maintenance, or repair requirements***, such as lengthening an inspection interval or changing the type of lubricant used in a compressor
- ***Site infrastructure changes***, such as fire protection, permanent and temporary buildings, roads, and service systems
- ***Organizational and staffing changes***, such as a reduction in the number of operators on a shift, a change in the maintenance contractor for the site, changing from 5-day operation to 7-day operation, or rotation of plant managers
- ***Policy changes***, such as changes in the amount of overtime permitted
- ***Other PSM system element changes***, such as modifying the MOC procedure to include a provision for emergency change requests
- ***Other changes***, including anything that "feels" like a change but does not fit in a category established for a facility; this "other type" should be in every MOC system

MOC system designers should (1) consider such example changes when developing the site's MOC system and (2) develop a similar list of changes and RIKs specific to the site. MOC system designers should also (1) consider the items in Table 3.3, (2) evaluate the frequency, sources, and types of change that are prevalent in their facility, (3) select the categories that make the most sense, and (4) consider including an "other" category to encourage workers to identify an MOC even if it doesn't seem to fit any of the established categories. Such a list is useful for training site personnel and as a reference for originators and designated initial reviewers. However, the ultimate scope of an MOC system should be a function of regulatory requirements and local needs. The examples provided in Appendix A may not apply to every facility.

Some companies have additional management systems that may apply to one of the types of changes above, and they may rely on that specific system rather than the MOC system for controlling those specific classes of changes (e.g., staffing changes, procedural changes). If certain classes of changes are controlled outside of the MOC system, these approved exceptions should be documented and carefully controlled. Also, the MOC training for site personnel should explain the basis for these exceptions.

3.4.3 Boundaries and Intentional Overlaps with Other Elements

MOC is only one system among many PSM practices likely in place at facilities. As shown in Table 2.3, MOC interacts with several typical PSM elements, using inputs from some elements and providing outputs to others. In addition, MOC may be used in situations other than process safety (e.g., environmental, security, quality). Also, other non-process safety-related management systems and administrative systems may exist that the MOC process must interact with (e.g., capital project management, budgeting, procurement). MOC interrelationships with other management practices should be well defined and understood to avoid accidentally omitting activities or unnecessarily duplicating effort.

For example, the primary output of an MOC system is the approved change for implementation. Other PSM elements (e.g., safe work practices, mechanical integrity) normally carry out the implementation. Process safety information (PSI) must often be updated based on the change, and the responsibility for achieving this can belong to the MOC process or to other management systems (e.g., operating procedures). Sometimes, however, companies find that updating documentation in a timely fashion is problematic, and subsequent audits reveal that the needed updates were not done. In those cases, companies sometimes decide to design overlap into the MOC system and the other related systems, whereby the MOC system checks to make sure that the follow-on work was performed by another appropriate management system before MOC closeout takes place (e.g., PSSR).

Mapping these overlaps and preventing gaps are important to ensuring MOC performance and efficiency. These activities should be addressed when defining the scope within the MOC system design specification.

3.5 INTEGRATING WITH OTHER PSM ELEMENTS AND EXISTING COMPANY PRACTICES AND PROGRAMS

An MOC system is a critical part of a company's overall PSM program. In fact, the MOC system provides many inputs to other components of a site's PSM program and helps ensure that all changes are appropriately addressed in other PSM policies, activities, and documentation. For this reason, the MOC system's design specification should define anticipated interfaces with other PSM elements and other administrative and management systems. In addition, the MOC system's developers should coordinate their efforts with site personnel who are responsible for carrying out the requirements of other company system guidelines. Some of the most important likely interfaces with PSM and other systems are described below.

Documents Created by Other PSM Elements

Documents from all appropriate PSM elements need to reflect the changes authorized by the MOC system. Therefore, the MOC system should provide descriptions of all relevant changes to other PSM elements in a timely fashion so that their information can be kept up to date. Some companies consider the interface between MOC and other PSM elements to be so important that they include, in the MOC system, verification that all appropriate documents were updated. For example, some companies include a step for verifying that piping and instrumentation diagrams (P&IDs) and procedures have been updated as a formal part of their MOC system, rather than including this step as a part of other management systems (e.g., PSI, operating procedures). Confirming that necessary documents have been updated may be established as a prerequisite for authorization to implement the change.

Companies should consider using a formal document control system to help manage the changes made to drawings, procedures, policies, and other documents. Document control may be managed through a paper-based or electronic system. Such a system can provide a directory of master documents and controlled copies, where appropriate. An effective document control system will (1) support MOC activities at the site, (2) provide reliable access to current PSM documentation, (3) establish a baseline for managing changes, and (4) prevent obsolete versions of controlled documents from remaining in circulation.

Process Hazard Analysis or Other Risk Studies

MOC documents can help a process hazard analysis (PHA or risk assessment team recognize new hazards or new safeguards for a process unit. This is particularly true for teams performing PHA revalidations based on previous studies. Therefore, records of all changes made since a previous PHA should be readily available to PHA teams as they update and revalidate a study. In addition, PHAs are a frequent source of proposed changes in a facility because of the action items that emerge.

Employee Training

Immediate training is often required for employees who will be directly affected by a change (e.g., training operators in a new procedure for regenerating catalyst). Furthermore, the MOC system should provide personnel who develop and conduct training with a description of all changes that affect training information or programs so that training documentation and programs can be kept up to date.

PSSR or Operational Readiness Reviews

MOC systems should be coordinated with PSSRs or operational readiness reviews to ensure appropriate coverage of changes of all types and sizes. Except for PSSRs conducted before initial startup of a new unit, all PSSRs could be initiated from the MOC system. Many companies combine their PSSR and MOC programs to ensure a fully integrated approach for safely resuming process operations following a change. Companies should consider conducting PSSRs for extensive changes and having the MOC system itself satisfy the PSSR requirement for smaller changes.

Incident Investigations and Compliance Audits

The MOC system should provide auditable records for use during incident investigations and compliance audits. Incident investigations may need to examine MOC records in order to determine the underlying causes of incidents or near-miss events. Compliance audit teams will need to examine MOC records to assess the effectiveness of the MOC system and make recommendations for system improvements. In addition, incident investigations and compliance audits are frequent initiators of change in a facility.

Other Facility Management Systems

Other systems currently in place to initiate or manage changes may also need to be modified or replaced to accommodate MOC. The MOC design specification should identify procedures and documents that may need modification or replacement, including procedures and documents associated with the following:

- New capital projects
- Maintenance work orders
- Instrument change requests
- Spare parts control, warehousing, and distribution
- Purchase requisitions
- Engineering change requests
- Research and development (R&D) recommendations
- Company specifications (e.g., equipment, products, raw materials, packaging)
- Personnel transfers
- Programming change requests
- Process experiments or tests
- Contractor service agreements (e.g., maintenance, engineering design, sourcing)

3.6 REQUIREMENTS FOR REVIEW AND AUTHORIZATION

Within its MOC design specification, management may want to specify requirements detailing the disciplines, departments, and organizational levels required to review and authorize different types of changes. Traditionally, many companies define authorization and review levels for project approval based on project cost. Similarly, the MOC design specification should recognize that some types of changes require more or less review, based on potential process safety variables (e.g., complexity of the change, magnitude of the change, hazards of chemicals and/or equipment involved).

A more sophisticated approach is to define the specific level of review required based on an assessment of the hazard and the likelihood of incidents that could result from the change. The MOC reviews and approvals may be parallel to or in series with traditional economic reviews and approvals for process modifications.

In addition, a company may wish to define the qualifications for personnel who are designated as official MOC reviewers or authorizers. These qualifications can be used for upgrading personnel for participation in the MOC system and also for designating personnel to serve as backups and temporary substitutes for those involved in MOC activities (e.g., vacation coverage, turnaround overtime).

3.7 GUIDELINES FOR KEY MOC ISSUES

Within its MOC design specification, management should identify the key issues and special situations they expect the development team to consider. Table 3.4 lists such key issues. See Chapter 4 for a discussion of the information that the development team should consider for these areas.

3.8 MAKING AN MOC SYSTEM EASIER TO MONITOR

An MOC system is typically one of the more active management practices in a PSM program, consuming significant resources. MOC sometimes requires a culture change on the part of employees in order to be effectively implemented. These issues require a company to be able to conveniently monitor the performance and efficiency of the MOC system. This monitoring has historically been performed via audits that are conducted every few years. Recently, an increasing number of companies have been establishing metrics for MOC systems in order to maintain "fingertip" control of the MOC process. The design stage of MOC is the best time to incorporate the means to easily monitor the MOC system.

TABLE 3.4. Key Issues to Resolve When Designing an MOC System

- System design capacity requirements
- Determination of an RIK verses a change
- Classification of the significance of the change
- Process or business need and the technical basis justification for the change
- Timing of the change
- Duration of the change
- List of information needed to review the change
- Checklists for ensuring that all elements of the change are addressed
- Expertise needed to review the change type
- Preferred hazard evaluation techniques for analyzing safety and health implications
- Tools available to the change reviewers
- Hazard/risk control/tolerance guidelines
- Documentation needs, forms, and retention policy
- Means for communicating changes to affected personnel in a timely fashion
- Means for providing employee awareness training
- Methods for achieving closeout of the MOC
- Whether to allow temporary changes, and identification of special conditions that should be associated with such changes
- Whether to allow emergency changes that circumvent part of the normal MOC system in order to accommodate urgent needs for change
- Variance/exception policy for special situations
- Means for monitoring and auditing the MOC system

3.8.1 Designing an MOC System to Make It Easier to Audit

Several design features can enhance a company's ability to efficiently audit its MOC system:

- Unique identifiers for RFCs
- Document retention policy for MOC records
- Policy of retaining previous versions of changed process safety information
- Location of MOC records (central or distributed)
- MOC summary spreadsheets or logs
- Electronic or paper MOC documentation

These features/capabilities should be addressed at the MOC design specification stage to ensure that the necessary MOC documentation is easily retrievable when audits are conducted.

3.8.2 Collecting Performance and Efficiency Measurement Indicator Data

Chapter 6 covers the use of metrics in continuous improvement of MOC systems. In order to have these metrics available for that purpose, they must be defined and systems should be established to collect the input data for them. The chapter on MOC in *Guidelines for Risk Based Process Safety* provides a list of example performance and efficiency metrics.[8] These examples are repeated in Chapter 6 of this book. At the MOC system design stage, company management should consider which metrics to use so that the data collection system can be established as the MOC system is designed and implemented, rather than waiting to do so after the fact, when the nature of the MOC system design may not facilitate collection of the desired metric inputs.

4

DEVELOPING AN MOC SYSTEM

Once the company has established the design specification for a management of change (MOC) system, they should assemble an interdisciplinary team of personnel to develop a written MOC program. The written MOC program will be used to educate and train site personnel on the MOC procedures. In addition, the MOC procedures help ensure consistent interpretation and application of management's policy for controlling changes throughout the life of the site. In some cases, written MOC procedures are not only a practical necessity, they are also required by regulations.

Table 4.1 compares typical MOC design tasks with the corresponding development tasks involved in implementing the MOC design specification. In addition to meeting the requirements of the design specification, the development team should anticipate continuous improvement activities based on feedback from personnel using the MOC system and from MOC system auditors.

The following tasks are performed when developing an MOC system:

- Verify implementation context
- Identify potential change situations
- Coordinate the MOC system with existing site procedures
- Establish request for change (RFC) review and approval procedures
- Develop guidelines for key MOC issues
- Design MOC system documentation
- Define employee training requirements
- Consider how the MOC system may be modified
- Compare the MOC system with the design specification

TABLE 4.1. Comparison of MOC Design and Development Tasks

Design Task	Development Task
• Define MOC system scope (site areas and activities)	• Work within scope
• Define system terminology (e.g., RIK definition)	• Use defined terms; enhance definitions as necessary (e.g., provide site/process examples of RIKs versus changes)
• Define roles	• Assign specific detailed tasks and responsibilities
• Define interface considerations	• Develop interface/transition procedures between MOC and other PSM systems
• Specify review and authorization guidelines	• Develop review and authorization procedures
• Specify guidelines for special situations, such as temporary repairs/ installations, emergency changes, variance policy	• Develop special procedures for these situations
• Specify requirements for other key issues, such as hazard evaluation communication of changes, special approval of high-cost items (if not already addressed by another system), system documentation	• Develop procedures, guidelines, forms, and other documentation, as required

The following sections address each of the MOC system development tasks.

4.1 VERIFYING IMPLEMENTATION CONTEXT

The design, development, and implementation of an effective MOC system should be based on the company's perception of the risk associated with the processes to which the MOC system applies. In addition, the rate at which the MOC system is used (thus placing demand on facility resources) and the facility process safety culture can also influence the design and operation of an MOC system.

Higher-risk situations usually require more formal and thorough development of an MOC protocol (e.g., a detailed written program that specifies exactly how changes are identified, reviewed, and managed). Companies having lower-risk situations may appropriately decide to manage changes in a less rigorous fashion (e.g., a general change management policy that is implemented via informal practices by trained key employees).

Facilities that experience high demand for managing changes may need greater specificity in the MOC procedure and greater allocation of personnel resources to fulfill the defined roles and responsibilities. Lower-demand situations allow facilities greater flexibility in developing an MOC protocol.

Facilities with sound safety cultures might choose to have MOC procedures that are more performance-based, allowing trained employees to use good judgment when managing changes in an agile system. Facilities with an evolving or uncertain safety culture may require more prescriptive MOC procedures, more frequent training, and stronger command and control management system features to ensure disciplined MOC implementation.

4.2 IDENTIFYING POTENTIAL CHANGE SITUATIONS

Based on the design specification supplied by management, the team must (1) define what is and is not a change for the site and (2) identify the specific types of changes that will be covered under the MOC system. [Appendix A provides examples of replacements-in-kind (RIKs) and changes for typical classes of changes that a company should consider when developing its MOC system.] The team should describe classes of changes using terminology that all personnel can understand. The goal is to provide a reasonably comprehensive list of change situations that the team expects could occur at the site. The list can be generated by several methods:

- Brainstorming among development team members (and possibly among groups of employees representing specific areas of expertise and responsibility, such as maintenance planners or operations supervisors)
- Reviewing existing procedures at the site to identify changes that are currently being managed through these procedures (e.g., work request)
- Reviewing previous incidents and near misses to identify where improperly reviewed or missed changes were causal factors
- Reviewing MOC systems and change procedures from other company locations as well as publicly available information[20]
- Discussing MOC strategies with specialists from other locations or similar companies

Sites subject to specific regulations [e.g., the Occupational Safety and Health Administration (OSHA) process safety management (PSM) rule] should refer to these documents to ensure adequate coverage of all of the MOC requirements included in such regulations.

Facilities making their first attempt to establish an MOC system should ensure that the development team carefully considers the workload implications of the classes and types of changes covered by the MOC system. While it may seem better to have the MOC system process more RFCs than may actually be required (or needed), facilities should be careful not to overburden the MOC system and its participants.

Once a representative list of potential changes at the site is created, the next step is to organize the list into categories. Each category of changes is defined such that changes in that category will require the same type of review and/or approval. This categorization is influenced by management's specification of those authorized to approve certain types of changes. The team should consider the following factors when performing the categorization:

- Departments and individuals who must implement the change (e.g., electricians versus pipe fitters)
- Departments and individuals who have particular expertise pertinent to the change being recommended (e.g., inspection department, rotating equipment experts, plant technical personnel)
- Departments and individuals affected by the change (e.g., operations versus maintenance)
- Departments and individuals who have authority over the entity being changed (e.g., operations for Unit #1 versus operations for Unit #2)
- Type and severity of the hazards associated with the change (e.g., changes to equipment in potable water service versus changes to equipment in hazardous chemical service)
- Special circumstances associated with a proposed change (e.g., temporary changes; changes during emergencies, off-shifts, holidays, the absence of key individuals)
- Staff groups that might be impacted by the change (e.g., safety, health, environmental, quality, security, information technology, human resources, purchasing, logistics)

The goal of this task is to establish a standard set of review and approval protocols for specific types of changes and circumstances that are likely to occur at the site. This standardization should help ensure that proposed changes receive timely, appropriate, and consistent reviews for approval, while minimizing the burden on the administrators of the MOC system.

4.3 COORDINATING THE MOC SYSTEM WITH EXISTING PROCEDURES

Numerous other management systems interface with a company's MOC system. The MOC system development team should consider how the MOC procedure would interact with each of these other administrative programs.

4.3.1 Maintenance Work Orders

For facilities that use a maintenance work order system, integrating MOC into that system provides an excellent mechanism for controlling facility change. Personnel involved in every aspect of the work order system (e.g., requesting work to be done, planning maintenance, approving work orders, implementing approved work orders) need to participate in managing changes. Otherwise, if the work order system does not adequately address MOC, the work order system may provide a way for uncontrolled changes to be made.

4.3.2 Spare Parts Control, Warehousing, and Distribution

MOC procedures should address spare parts control because it represents the potential for inadequate material to be used, even when personnel intend to comply with MOC procedures. Companies often find that they have to upgrade procedures for receiving, inspection, material labeling, storage, and spare parts inventory control to prevent RIKs from becoming unauthorized/unintended changes.

4.3.3 Purchase Requisitions and Suppliers

Purchase specifications and requisitions are an important link in the chain of documents required to obtain and install equipment in process systems. Controlling changes in purchasing specifications and requisitions is one area that an MOC development team should consider when designing MOC procedures. The extent to which all changes in purchase requisitions are reviewed depends in part on how the scope of the site's MOC system was defined. (That is, are only some purchases covered in the scope of the MOC system?) In addition, some companies control changes in suppliers of material, equipment, or chemicals even though the purchase specification for the item has not changed. This control helps avoid problems created by differences between specific suppliers (e.g., differences in trace contaminants, packaging, delivery method).

4.3.4 Engineering Change Requests

Engineering change requests, or similar design control mechanisms, provide a way for personnel to request that engineering effort be authorized to consider changes in a facility. The MOC system development team needs to determine at what point (or points) changes that proceed through this formal mechanism will be reviewed to satisfy the MOC requirements.

4.3.5 Research and Development Recommendations

Research and development (R&D) recommendations are often used to improve process systems or test new ideas. However, they are often made by personnel who are not closely involved in the process system operation. Therefore, careful review via the MOC system of R&D recommendations by operations, maintenance, process engineering, and environmental, safety, and health (ESH) personnel will ensure that they do not have impacts that were not envisioned by R&D personnel.

4.3.6 Company Standards and Specifications

Company (or licensor) standards and specifications may provide a mechanism to control material in specific process systems, engineering and construction practices, and other areas. However, because such documents often apply to many different kinds of facilities, additional review of changes in company standards and specifications at the local level helps ensure that the changes are appropriate for the specific process applications at that location.

4.4 ESTABLISHING RFC REVIEW AND APPROVAL PROCEDURES

For each category of change, the development team determines the steps (i.e., reviews, actions, and approvals) that will be required before a change is implemented. In making these decisions, the team should use as a guide the criteria provided by company management concerning the level of authority required to approve certain types of changes. However, for each type of change, the development team should consider including the following five key steps in their MOC procedure (see Table 4.2).

Initial review. The initial review step considers whether a change is necessary and whether it is truly a change based on the definition developed for the MOC system. If the proposed change is an RIK and not a change, it can be implemented without any further review. The RFC form should be returned to the RFC originator so that the originator knows the action can be implemented. However, the form (or a copy of the form) should be filed with other MOC documentation. This allows the decisions of the initial reviewer to be audited and requirements for the implementation of the MOC review to be revised, if appropriate. Some companies also use the initial review to stop proposed changes that are not aligned with long-term facility goals, are unaffordable or infeasible, or are otherwise not likely to be supported by management. This saves time and resources that might otherwise Sbe devoted to other, more viable change requests.

TABLE 4.2. Key Steps in an MOC System

Step	Focus
Initial review	Is the proposed change necessary? Based on MOC system definitions, is it a change? Is it covered by another procedure or management system?
Classification review	Is the change extensive or complex enough to require a multidisciplinary review? Who needs to review the change?
Hazard review	Have potential problems been identified and have required controls been documented?
Authorization review	Have all identified hazards and associated tasks required prior to implementation been addressed and documented?
Closeout review	Have all identified hazards and associated tasks required after implementation been addressed and documented?

Classification review. The classification review step determines who needs to perform the subsequent reviews. For example, one person may be able to adequately handle the entire hazard review step for some types of changes, although many facilities require at least two reviewers for any change. For other types of changes, most facilities require several people or departments to review and approve the change. The approvals may involve all of the organizations that are directly affected by the change and the organizations that conduct the reviews required by the change. Specific levels of authority for approving changes within each organization should be defined for different types of changes (e.g., some changes require a higher level of authority based upon their safety significance). *Remember, designating approval authority at too high a level may impede the MOC process, possibly encouraging people to bypass the system to "get things done." However, designating authority at too low a level may lead to ineffective change control.*

The development team should determine which technical activities are necessary for each class of change. However, defining standard MOC procedures for every conceivable change situation is not practical. For proposed changes that do not have clearly defined standard review procedures (i.e., those changes that do not fit within the established RFC categories), the MOC coordinator (or a designated substitute) should establish review requirements on a case-by-case basis. The development team should provide some general guidance for specifying reviews in these special circumstances. As new types of changes are reviewed in this manner, the changes may be classified within existing categories (if appropriate), classified in a new category, or treated as an isolated case that does not need to be included in the standard MOC procedures.

Hazard review. Regardless of who performs the hazard review step, the objectives are the same. The main tasks performed by the reviewers are as follows:

- Identify the hazards introduced or exacerbated by the proposed change. The reviewers may request that a team conduct a more extensive hazard evaluation [e.g., formal process hazard analysis (PHA) or risk assessment].
- Determine whether the change can be implemented safely (this should consider both process safety concerns and traditional safety concerns, such as industrial hygiene and personal protective equipment).
- Determine whether the proposed controls under which the implementation is to be made are adequate (e.g., design features, special permits, additional staffing, specific supervision required during implementation).
- Determine the additional activities that must be accomplished prior to implementation of the change (e.g., updating process safety information developing operating procedures, determining the level of personnel

training or at least communicating the change to personnel, purchasing associated material).

- Determine whether other reviews are required. The MOC reviewers may require that other disciplines (e.g., environmental, electrical, mechanical) review the change prior to its implementation. This decision can be reserved for situations in which the MOC reviewers feel that they do not have the expertise needed to ensure that the change is adequately reviewed. Some companies always implement a broader, safety, health, and environmental review. These companies find that such a review can satisfy a number of regulatory and management purposes (e.g., impact on environmental permits, changes to risk management plans).
- Determine the appropriate depth of a pre-startup safety review (PSSR) or operational readiness review (ORR), if required.
- Identify the actions that need to be accomplished and documented after the change is complete to satisfy regulatory or company requirements (e.g., preparing as-built drawings).

Authorization review. The authorization review step serves as a final MOC approval mechanism prior to implementation. This review should ensure that the actions required prior to implementation of the change, based on the hazard review step, are complete and properly documented.

Closeout review. The closeout review step serves as the final MOC review of the change. In this step, the reviewer indicates that all of the post-implementation activities are complete. Such activities may include updating and issuing revised drawings, filing PSSR documentation, and other activities required by the hazard review step but not necessary prior to implementation of the change. Historically, post-implementation activities have often been neglected. However, the MOC system should track changes through to completion via the closeout review step and ensure that activities are completed in a timely manner.

Figure 4.1 is an example flowchart for a simple MOC system.

4.5 DEVELOPING GUIDELINES FOR KEY MOC ISSUES

The development team should create specific guidelines to help MOC system users address some of the key MOC issues, such as evaluating hazards, communicating changes, tracking temporary changes, integrating MOC with ORRs and PSSRs, and allowing emergency changes.

4.5.1 Evaluating Hazards

An important aspect of MOC reviews is assessing the hazards associated with proposed changes. This primarily includes traditional hazard evaluations, but it can also include evaluating the hazards associated with physically implementing the change (e.g., hazards to personnel installing new equipment). The MOC development team should determine the level of hazard evaluation needed for specific types of changes. This could include describing (1) the scope of the **hazard review** step, (2) the level of detail needed, and (3) the specific issues that must be addressed, as well as suggesting some appropriate hazard evaluation techniques. Checklists of questions are often used to (1) promote critical consideration of hazards associated with a proposed change, (2) assist in evaluating hazards, or (3) assist in determining the depth of review needed.

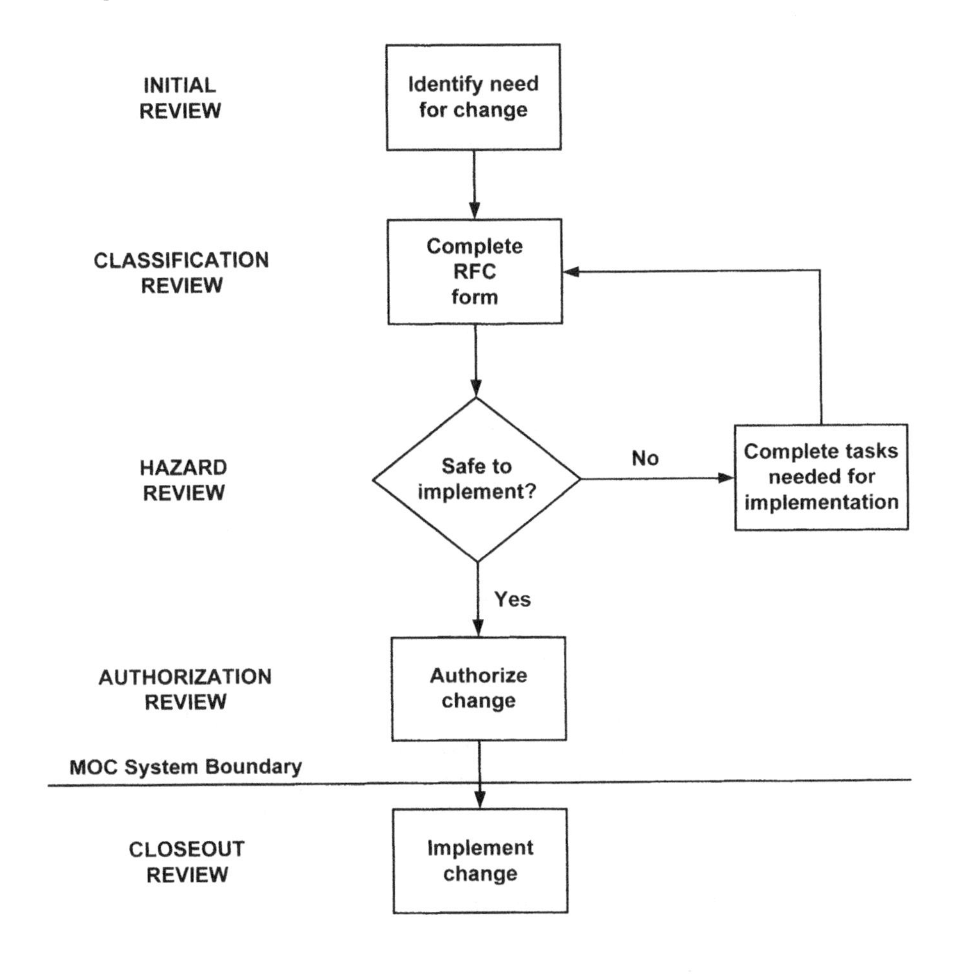

FIGURE 4.1 MOC System Flowchart

Site management may decide that formal hazard evaluations are necessary for certain types of changes. The MOC system should have formal criteria for initiating these analyses (although the administrators of the PHA system should have the freedom to define the most appropriate hazard evaluation techniques for the required analyses on a case-by-case basis). Also, the MOC system should provide sufficient information about the change to conduct a hazard evaluation. *Guidelines for Hazard Evaluation Procedures, Second Edition with Worked Examples* provides detailed descriptions of hazard evaluation techniques and their strengths and weaknesses.[17]

4.5.2 Communicating Changes or Providing Training

Communicating changes or providing appropriate training to affected employees is an essential element of an MOC system. MOC systems should contain specific methods for ensuring that change information is communicated to affected employees in appropriate detail before the change is implemented or prior to the employee's work shift during which he or she will be affected by the change. In addition, some changes may involve significant revision to the operating or maintenance practices employed such that simple awareness communication will not be sufficient. In these situations, affected employees may require detailed training on the revised procedure/practice. Such communication/training methods may include verbal communication from supervisors, formal training sessions, change notices documented in procedures, entries in logbooks, written summaries of changes, e-mail notifications, and other approaches. Some companies use (1) daily pre-shift safety briefings to discuss recent changes and (2) monthly safety meetings to communicate change information that may not have an immediate impact on site personnel.

A site may need more than one method for communicating changes, depending on the attributes of the specific change (e.g., magnitude, associated hazards, urgency, number and types of departments involved). Whatever the means used to communicate change information, the training should focus on how the change affects both the hazards of the process and the tasks performed by various individuals. Each site and/or process area should develop an efficient means of documenting the communication. This documentation may include verification or confirmation that affected personnel have been informed of or trained on the change prior to its implementation.

4.5.3 Tracking Temporary Changes

Temporary changes (e.g., jumpers/bypasses for instrumentation and control schemes, unavailability of process or safety equipment, trial use of a new piece of equipment) typically receive special consideration because the facility may accept a somewhat greater short-term risk for a predetermined, finite time than what is normally tolerable for long-term operation. For example, equipment that meets the specification may not be immediately available or the site is experimenting with a new design. Other temporary changes (e.g., clamps, patches, leaks stopped by injection of materials) receive special consideration because they may be designed for a short-term service life. In any of these cases, a temporary change is approved with the stipulation that the change last no longer than a specified time and that additional safety features be implemented (e.g., administrative or procedure changes to provide additional surveillance, additional instrumentation).

Companies should also be aware that the number of potentially affected personnel who should be informed of and/or trained on the change may increase if temporary changes are extended without further review. *Temporary changes need prescribed time limits, appropriate authorization, and proper documentation. The MOC system should have a mechanism for monitoring the status of temporary changes, ensuring that time limits and any other stipulations are not violated.*

Finally, the system should have a means to ensure that any changes to procedures or PSI that were made because of the temporary change are returned to normal when the temporary change is reversed.

4.5.4 Integrating MOC with ORRs and PSSRs

The Center for Chemical Process Safety's *Guidelines for Risk Based Process Safety* framework has defined ORRs as an enhanced element, incorporating the traditional PSSR concept. The main difference between the traditional OSHA PSM compliance-based PSSR element and the readiness element is that ORRs are to be performed whenever the process/activity has been out of service or in an alternate or idle configuration longer than allowed in the normal operating procedures for the process/activity. PSSRs are a subset of all ORRs that may be performed by a facility. Some ORRs may be conducted following an MOC review, so considering the integration of MOC and readiness activities is important.

Major changes to a process (e.g., those involving changes in the design documentation and operating procedures for a process, such as major equipment additions, modifications, or deletions) often require PSSRs to be conducted before the process returns to operation. The PSSR confirms that (1) the equipment is in accordance with design specifications, (2) appropriate procedures are in place, and (3) the necessary training has been completed. Since these issues should be addressed as part of the MOC system, a PSSR serves as a final check of changes authorized by the MOC system. Such a check is important because major modifications often involve many changes, possibly resulting in an item or an issue being overlooked during MOC reviews.

Traditionally, MOC procedures have addressed changes in, or minor additions to, existing facilities, while PSSRs have been performed (1) for new facilities, (2) prior to restart of mothballed operations, or (3) after extensive modifications that have required shutdown of existing facilities and restart. This approach meant that MOC and PSSR programs did not generally overlap. However, facilities complying with the OSHA PSM regulation (29 CFR 1910.119) should have closely related MOC and PSSR programs.[3] Table 4.3 presents the specific OSHA MOC and PSSR requirements side by side.

Companies should coordinate their MOC and PSSR/ORR programs.[8,19] For each change, the MOC procedure could include a determination of the appropriate level of PSSR/ORR, in addition to performing the pre- and post-implementation actions identified in the MOC reviews. If a formal PSSR is not required, the MOC program should simply ensure that the basic PSSR questions were satisfied. In addition, MOC procedures should:

- Include measures to confirm that equipment is in accordance with design specifications
- Ensure that updating of procedures includes safety and maintenance procedures in addition to operating procedures
- Ensure that training is up to date

To coordinate MOC and PSSR procedures, companies may want to develop a single procedure that describes both of them. Or, to illustrate how they are effectively coordinated, companies should, at a minimum, ensure that the separate procedures for MOC and PSSR reference each other and explain how MOC and PSSRs work together to ensure that (1) all of the regulatory requirements are met and (2) changes, both minor and major, are adequately reviewed.

4.5.5 Allowing Emergency Changes

Some organizations have found it beneficial to develop an MOC procedure that can be implemented with minimal delay. Use of this type of emergency procedure, if considered necessary, should be restricted to situations for which the time required to implement the normal change procedure would not be acceptable (i.e., serious consequences could occur if the change is not made promptly). For example, one American Chemistry Council member company has defined an emergency as "a situation that requires immediate action to avoid equipment damage, personnel hazard, environmental violation, or severe economic penalty." *Use of an emergency procedure should be a relatively rare occurrence. An emergency procedure must not be used simply to avoid the work associated with implementing the normal procedure.* One way to discourage this practice is to require whoever uses the emergency procedure to follow up with a normal change request and satisfy all of the normal MOC requirements.

TABLE 4.3. Comparison of OSHA MOC and PSSR Requirements

Management of Change	Pre-Startup Safety Review
The employer shall establish and implement written procedures to manage changes (except for "replacements in kind") to process chemicals, technology, equipment, and procedures; and changes to facilities that affect a covered process	The employer shall perform a pre-startup safety review for new facilities and for modified facilities when the modification is significant enough to require a change in the process safety information
The procedures shall ensure that the following considerations are addressed prior to any change: • The technical basis for the proposed change; impact of change on safety and health • Modifications to operating procedures • Necessary time period for the change • Authorization requirements for the proposed change	The pre-startup safety review shall confirm that, prior to the introduction of highly hazardous chemicals to a process: • For new facilities, a process hazard analysis has been performed and recommendations have been resolved or implemented before startup; modified facilities meet the requirements for management of change • Construction and equipment is in accordance with design specifications
Employees involved in operating a process, and maintenance and contract employees whose job tasks will be affected by a change in the process, shall be informed of, and trained on, the change prior to startup of the process or affected part of the process	Training of each employee involved in operating a process has been completed
If a change covered by this paragraph results in a change in the process safety information required by paragraph (d) [of 29 CFR 1910.119, such information shall be updated accordingly	
If a change covered by this paragraph results in a change in the operating procedures or practices required by paragraph (0) [of 29 CFR 1910.119, such procedures or practices shall be updated accordingly	Safety, operating, maintenance, and emergency procedures are in place and are adequate

The MOC development team should (1) define the circumstances under which the emergency procedure can be implemented, (2) develop requirements that focus on quickly evaluating the safety of the immediate situation, and (3) require that the remaining MOC requirements be completed shortly after the change is implemented. Such procedures generally require the involvement of personnel who are always available (e.g., personnel assigned to the shift) or some form of call-out procedure. One key MOC principle that should be maintained even in emergency situations is involving several people who represent different disciplines, if possible.

4.6 DESIGNING MOC SYSTEM DOCUMENTATION

The development team should establish the documentation format for the MOC system, which should:

- Describe the proposed changes
- Establish the required reviews, actions, and approvals
- Document approval(s) for changes
- Track the status of temporary changes
- Provide summaries of actual changes to affected organizations and individuals

Some of these needs can be met with one form or document. Appendix C contains an example RFC form with spaces provided to describe a change, specify the required MOC reviews/approvals, and document approvals of a change. The completed form can serve as a summary of the change.

In designing the documentation system, the development team should include policies for records retention for the various types of documentation. Finding corporate document retention requirements to be in conflict with relevant regulatory guidelines is not uncommon; such conflicts will need to be resolved.

In addition to documenting a particular change, an MOC system document should be available that describes the system and its procedures. Such a document (commonly called a written program) is required by the OSHA PSM regulation and other accident prevention regulations (e.g., state laws, the Environmental Protection Agency's Risk Management Program rule). The written program can be used as a training manual and as a user's manual for the MOC system and should define the requirements that the MOC system is likely to be audited against.

4.7 DEFINING EMPLOYEE TRAINING REQUIREMENTS

Developing and conducting training for all personnel involved in MOC is critical to the success of a new or revised MOC procedure. *Many systems have failed or at least encountered severe problems because personnel did not understand why the system was necessary, how it worked, and what their role was in its implementation.* The development team should define how each type of training would be developed and provided, including:

- Awareness training to educate all affected personnel on how to recognize changes within the scope of the MOC system (e.g., maintenance craft personnel who should inquire about the MOC before installing a change)

- Initial training for personnel who will be involved in the MOC system (e.g., personnel who are expected to request changes or provide initial reviews)
- Initial training for personnel who will have major roles in the operation of the MOC system (e.g., reviewers at other levels and approvers), addressing topics such as hazard evaluation methods, use of specific technical analysis tools, etc.

Refresher training for all MOC system personnel should also be considered. Maintaining examples of problems both avoided and created by MOC for use in such training is often helpful. Appendix A has some examples that may be useful in training; however, using feedback from a site's own system is most effective. Unfortunately, feedback provided by incident investigation reports is more often available than information about incidents that are avoided. Section 5.5 provides advice on topics to include in general MOC awareness training.

4.8 CONSIDERING HOW TO MODIFY THE MOC SYSTEM

MOC systems are one of the most frequently upgraded management system procedures. The development team should consider how to address suggested changes to the MOC system. The approach should address (1) ways in which personnel can propose changes to the system, (2) reviews and approvals needed for implementing a change, (3) methods of communicating changes to personnel involved in the MOC system, and (4) methods for updating MOC system documentation.

4.9 COMPARING THE MOC SYSTEM TO THE DESIGN SPECIFICATION

The development team and senior management need to review the MOC system to ensure that it meets the requirements established by regulations and management. Furthermore, the development team and senior management should make certain that the MOC system is understandable to potential users and convenient for them to use. Many companies recommend having all organizations involved in MOC procedures review and approve the MOC system prior to its first use. Developing a flow chart (such as the one in Figure D.2 in Appendix D) provides a visual representation of the proposed MOC system and facilitates this type of review.

5

IMPLEMENTING AND OPERATING AN MOC SYSTEM

Before implementing the management of change (MOC) system, the development team should consider the following foundational activities to help ensure the success of the system:

- Preparing the site infrastructure to support MOC activities
- Managing the culture change
- Integrating the MOC system with existing site procedures
- Developing a phased implementation plan for the MOC system (including a field test of the system to identify problems)
- Training affected personnel on MOC procedures

A facility can then roll out the system, operate it, and maintain it. The implementation team should have representatives from typical facility departments (e.g., operations, maintenance, engineering, safety).

5.1 PREPARING THE SITE INFRASTRUCTURE

The MOC system is an integral part of the process safety management (PSM) system. If the PSM system or the site is not ready to operate an MOC system, then the hazards of changes will not be properly managed, records will not be kept up to date, and training will fall behind, potentially undermining process safety at the facility.

The MOC implementation team should prepare the site for efficient startup and operation of the MOC system. Table 5.1 lists some issues that may need to be addressed prior to MOC rollout.

5.2 MANAGING THE CULTURE CHANGE

MOC systems keep people from making changes without the appropriate review and approval. In most cases, these changes are intended to either (1) improve operability or (2) sustain operations affected by equipment failures or external events. As a result, some people consider MOC systems to be impediments or barriers to getting work done. If that attitude is prevalent, company management should take steps to address this culture prior to or coincident with implementation of a new MOC system.

Sometimes the culture shock can be lessened by involving people from various disciplines, departments, or perspectives in the MOC design, development, troubleshooting, and solution development activities. Efficiency in design of the MOC system can be made a priority to ensure that the system is "just big enough" for its intended use. Another way to manage culture shock is to ensure that site personnel understand (1) the reasons the MOC system is being deployed, (2) the importance of having a healthy MOC system, (3) the

TABLE 5.1. Issues that Should Be Considered Prior to MOC Rollout

- Confirming management commitment and resources
- Assembling up-to-date PSI
- Preparing all the tools and forms needed to execute MOC activities
- Securing resource commitments from infrastructure departments (e.g., information technology)
- Preparing all MOC support materials (e.g., spreadsheets, files)
- If using an electronic MOC system, establishing a fully tested system with sufficient network resources dedicated to its maintenance
- If using an electronic MOC system with contract personnel involved, obtaining approval to provide computer access and e-mail accounts to contract employees for MOC notification
- Having completed, refined, pilot-tested MOC procedures
- Establishing a means for collecting MOC metric data and other feedback
- Developing a plan for conducting MOC management reviews
- Drafting plans for auditing the MOC system
- Establishing a means for communicating changes and informing/training potentially affected personnel
- Developing a plan for MOC rollout training

potential (and case histories of) consequences of not having a good MOC system in place, and (4) management's expectations for MOC conformance. If one facility's existing MOC system is being implemented at another facility, personnel associated with the existing system can be brought in to champion the MOC system at the new site.

Consideration of facility culture should be factored into the implementation plan and MOC awareness training. Should MOC culture problems persist (as evidenced by metrics, audit results, incident root causes, or management reviews), identifying the underlying causes of the cultural issues and implementing corrective actions will be necessary. See Section 6.4.2 for a discussion of addressing cultural issues.

5.3 INTEGRATING THE MOC SYSTEM WITH EXISTING PROCEDURES

Although the development team considers existing site procedures when creating the MOC system, the implementation team may also need to review the MOC system to help eliminate conflicts with existing procedures (e.g., work order system, capital project management system) during the installation phase. Any conflicts or overlaps should be resolved by modifying the existing procedure or the MOC system. The MOC system will be more successful if it complements existing procedures.

5.4 DEVELOPING A PHASED IMPLEMENTATION PLAN

When possible, the development team or implementation team should plan a phased implementation, including a field test of the MOC system procedures and documentation in a selected part of the facility to identify and correct any weaknesses. *No substitute exists for testing the MOC system on actual change situations in the facility where it will be used.* Phased implementation allows the implementation team to evaluate the training and monitor the startup of the MOC system in selected process areas before full-scale rollout and operation. This phased implementation provides additional opportunities to fine-tune the program for efficiency and effectiveness. Once the program is proven in one area, the scope of the program can be expanded to cover other process areas as required.

The field test or initial phase of implementation should address all of the major features of the MOC system, including special circumstances, such as temporary and emergency changes. The team can record and summarize the test data and present the results to site management for discussion. The duration of this field-testing varies, but it may require several months, depending upon how many changes the area experiences. The implementation team may need to revise the MOC system based on the results of the field test.

Where phased implementation is not possible (e.g., expedited implementation is needed for compliance reasons), the implementation team should conduct tests using simulated changes of various types before implementing the MOC program. Be aware that implementing an untested program can be detrimental. Personnel frustrated by a poorly implemented program may lose confidence in the MOC system and try to bypass part or all of it.

5.5 TRAINING PERSONNEL AFFECTED BY MOC PROCEDURES

This section provides examples of topics to include in awareness training, which is conducted to educate personnel about the MOC system. Companies can use these examples, generate their own context-specific topics, or employ a combination. This information may also be useful for detailed MOC training. However, detailed training will likely focus on site-specific MOC procedure issues.

All personnel involved in making changes at the site (including contractors, if the system is so designed) should be trained on the MOC philosophy, procedures, and documentation. This training should include workers at all levels of authority, with special emphasis on the line maintenance and operating personnel, as well as frontline supervisors. Special emphasis on these employees is necessary because they are often the key to identifying, describing, and classifying change situations in day-to-day operations. The awareness training should emphasize practical examples that personnel can use as guidelines in day-to-day operations.

MOC system awareness training should cover the following issues:

- The importance of MOC procedures and the general MOC philosophy.
- Definitions and terminology associated with the MOC system should be introduced
- Company or industry case histories citing MOC failure as a contributing factor
- Regulatory or legal obligations relating to MOC
- How the MOC system interfaces with existing procedures (e.g., work order system)
- General roles and responsibilities of employee groups or individual employees
- How to recognize change situations [especially differentiating between replacements-in-kind (RIKs) and actual changes, as well as recognizing the subtle changes that can occur]. Checklists to aid identification of change situations could be helpful, and lists of anticipated change situations that have already been evaluated can improve performance
- Examples of typical and unusual changes

- How to initiate an RFC
- How to classify a change for MOC review (particularly under special circumstances)
- How the MOC review process works (especially the responsibilities of personnel who must conduct specific reviews and issue approvals)
- Overview of hazard evaluation concepts and techniques
- How the MOC system is documented
- Special features of the MOC system, particularly temporary and emergency changes
- How to recommend changes to the MOC system
- Who can answer MOC questions

To be effective, MOC system awareness training cannot be a one-time activity. Changes to the MOC system require updating the training for employees who participate in the MOC process. Any new personnel (or personnel who transfer from an area that does not use the MOC system) should receive MOC system awareness training as appropriate. Also of value is ongoing training provided through periodic workshops covering key MOC issues and experience with the MOC system. Finally, the MOC coordinator is likely to provide a coaching, counseling, or training function on an ongoing day-to-day basis.

5.6 OPERATING AN MOC SYSTEM

The MOC coordinator directs the operation of the MOC system by (1) monitoring the operation of the system, (2) resolving questions and disputes relating to the system, and (3) maintaining system documentation and records. Some sites may need to provide the MOC coordinator with clerical or other support to help administer the program and manage MOC documentation and recordkeeping. At smaller sites, MOC coordination may simply be a part-time responsibility for an employee already in an appropriate position. Whether the MOC coordinator is full-time or part-time, management should ensure that MOC responsibilities are covered when the primary MOC coordinator is not available.

At large sites, assigning an MOC coordinator to each operating area, who can address the required changes for that area on a day-to-day basis, may be beneficial. However, if MOC coordinators are distributed across different areas, the site-wide MOC or PSM coordinator should ensure that the MOC procedure requirements are consistently applied.

5.6.1 Monitoring the Operation of the MOC System

The MOC coordinator ensures that (1) the phased implementation of the MOC system is progressing in a timely manner, (2) the MOC system is working well with other PSM systems and other site procedures, (3) the MOC procedures are being followed as intended, and (4) the MOC system is fulfilling its design intent. The MOC coordinator monitors these factors through routine administration of the MOC system and periodic internal audits/reviews of the MOC system (see Chapter 6). Any deficiencies in the MOC system are corrected according to the procedures for modifying the MOC system. Deficiencies in other systems or activities performed by other organizations should be resolved by the MOC coordinator in conjunction with the managers of specific areas or with the PSM manager.

The MOC coordinator should consider monitoring emergency and temporary changes closely. Any changes implemented on an emergency basis must also be reviewed in more detail after their implementation to ensure that the full MOC protocol is implemented. Also, if simplified procedures are defined for making emergency changes, the potential exists for abuse of the emergency change status. Examining the reasons that emergency changes were processed allows the MOC coordinator to (1) revise the MOC procedure to reduce the need for emergency changes, (2) re-educate those who do not understand the design intent of the MOC system, or (3) focus management attention on organizations or individuals who abuse the MOC system.

5.6.2 Resolving MOC Questions and Disputes

Individual departments may be largely responsible for implementing the MOC procedures without significant interaction with the MOC coordinator. However, when uncertainties in the interpretation of MOC procedures arise (e.g., a proposed change does not fit into any change category that has a prescribed review process), or when reviewers disagree about an MOC requirement, the MOC coordinator should resolve these issues. If a conflict cannot be easily resolved, the PSM coordinator (and possibly senior managers) should help resolve them. All of these issues should provide feedback to the MOC awareness training program and potentially to the MOC management review and audit activities.

5.6.3 Maintaining MOC System Documentation and Records

The MOC coordinator or the coordinator's designee is responsible for maintaining documentation for the MOC procedures, as well as the records of changes at the site.

MOC procedures documentation. The MOC procedures (including all tools, guidelines, and software) should be updated as needed so that personnel with responsibilities under the MOC system are always aware of current procedures. MOC procedures should be controlled so that the procedures can be readily updated as changes occur and the facility can ensure that the workforce has easy access to the current MOC procedure (e.g., via the facility intranet).

MOC records. The records of requested changes, change approvals, and tracking forms for temporary changes should be archived for use in monitoring the MOC system and for use by other PSM systems (e.g., process hazard analyses, compliance audits). The MOC coordinator should ensure that MOC records are retained in keeping with the site/company records retention policy.

MOC documentation can be paper based or electronic. Many companies are moving toward using computer networks and integrating MOC system documentation with existing work order, drawing, and procedure documentation systems. MOC software applications are addressed in more detail in Appendix D.

6

MONITORING AND IMPROVING AN MOC SYSTEM

Why would anyone want to improve their management of change (MOC) system? Why not just maintain the status quo? Within a process safety context, these seem like ridiculous questions, but industry examples of company behavior – judged after an accident – indicate that such concerns would have been justified and that improvements were badly needed. In this age of shrinking resources, asking such questions and expecting answers is reasonable.

6.1 MOTIVATIONS FOR IMPROVEMENT

Table 6.1 provides some observed industry motivations for improving MOC.

TABLE 6.1. Possible Motivations for Improving MOC
• Recent major accident
• Series of incidents
• Regulatory considerations (new rule or enforcement actions)
• Industry group membership obligation
• Peer pressure/comparisons of existing practices
• Perception that risk is not tolerable/increasing
• Resource pressures
• Desire to be more profitable
• Company policy of continuous improvement

Some of these motivations are positive (e.g., continuous improvement driven by a strong company commitment to quality). Most are more negative in nature (e.g., accidents with consequences), while others fall in the middle (e.g., peer pressure, risk perceptions). The public might believe that most companies interpret the business case for profitability and attractiveness to investors/suitors as justification for allocating optimum process safety resources over the long haul. However, anecdotal experience has shown that company objectives tend to focus more on shorter-term goals related to quarterly numbers. This is true despite significant efforts by the Center for Chemical Process Safety (CCPS) and various member companies to highlight the business case for process safety resource investment (e.g., *The Business Case for Process Safety*[21]).

Experience has shown that MOC is one of the more difficult process safety management (PSM) elements to implement – and to get and keep right! Typically, companies with long-standing MOC systems have revised their systems many times over the years as they have learned from experience (both good and bad). Some organizations continue to struggle with personnel who view MOC as an impediment to progress, and these organizations suffer the consequences of uncontrolled changes as the MOC system is continuously circumvented. Other organizations may have tight control over changes, but they are disconcerted by the amount of time and resources being consumed by efforts to implement and administer the MOC system.

Chapter 2 introduced the concept of Risk Based Process Safety (RBPS) and MOC effectiveness, which was defined as a function of both performance (i.e., achieving the right results) and efficiency (i.e., achieving those results with the appropriate expenditure of resources). This chapter, however, provides guidance on how to diagnose and repair an MOC system that is broken, or to optimize an MOC system that is not working as effectively as the needs of the organization require.

Two main categories of MOC improvement activity exist:

1. Corrective action to fix a seriously deficient MOC system via redesign or reimplementation
2. Continuous improvement of a working MOC system using available effectiveness enhancement methods

Corrective action is the more serious and time-consuming activity. However, if the need is indicated by serious accidents, incident trends, or chronic MOC deficiencies highlighted by audits, then a site may have no choice but to either redesign the system or re-implement the original design (if a review confirms that the original design was suitable, but just not properly implemented).

In addition to the circumstances listed in Table 6.1, an organization may perceive other reasons to redesign an MOC system that is believed to be dysfunctional or at variance with the needs of the organization. Examples of such reasons are listed in Table 6.2. Multiple motivations for redesign are not uncommon, and the design solutions for each may not be the same. Consequently, company management should gain as much understanding as possible about the needs that should be addressed.

6.2 SOURCES OF INFORMATION TO LAUNCH AND GUIDE IMPROVEMENT

Fortunately, companies have a plethora of sources of information (see Table 6.3) to help launch and guide their MOC improvement efforts – if they only take the time to gather and analyze it.

Evaluating incidents and near misses can help identify excellent opportunities for MOC system improvement. Asking personnel (e.g., operators, operational managers, maintenance employees) about potential failures and operational concerns is another excellent way to help identify what is broken or likely to break. Peer group benchmarking and sharing best practices are higher-level sources of experience-based information that can help participants understand and address identified risks.

TABLE 6.2. Possible Reasons for Wanting to Redesign an MOC System

- Persistent MOC findings in audits and/or management reviews
- Personnel feedback
- Unfavorable performance or efficiency metrics
- A major change in technology or processes that modifies the facility risk perspective in a manner warranting more (or less) rigor in the MOC system
- A perception that the risk from unmanaged changes is increasing or, perhaps, is already intolerable
- A merger or acquisition that compels the facility to adopt the MOC protocol of the new owner or that stimulates the new owner to emulate the better MOC program of the acquired facility
- A change in company policy or a desire to standardize across the company
- Peer pressure (or greater peer awareness) as a result of benchmarking/comparing existing practices
- Resource pressures (e.g., MOC system requirements that currently exceed available resources or require some existing resources that are about to be cut)
- Significant site or corporate reorganization or restructuring, or the outsourcing of a key function (e.g., engineering design, procurement)
- The introduction of new industry guidance (such as this book)

TABLE 6.3. Sources of Information for Improving an MOC System

- Incidents, root cause analyses and investigation reports
- Performance and efficiency measures
- Financial indicators (e.g., losses, insurance costs)
- Introspective reviews (e.g., audits, PHAs)
- Sharing of best practices within industry groups
- Benchmarking within peer groups
- Global evaluation of state-of-the-art practices

Within a single facility, five main sources of data/information exist on which to base MOC system improvements:

- MOC audit results
- Collected MOC system activity data
- MOC performance and efficiency metrics
- Results of incident investigations
- Results of introspective reviews that identify MOC problems [e.g., process hazard analyses (PHAs)]

6.2.1 Performing MOC Audits

Companies should consider performing periodic internal and external audits of the MOC system to help ensure that the system's goals are being met. Internal audits are those sponsored by the MOC coordinator. In some cases, the MOC coordinator may involve other personnel when performing an audit (e.g., former MOC development team members). Supervisors or managers who are familiar with the MOC system can also assist, as long as they are not routinely involved in the request for change (RFC) review/approval protocols for the area being audited.

External audits are those conducted by company personnel who are not associated with implementation of the MOC system, or by qualified third parties. These individuals usually perform the audit under the direction of the site or corporate PSM manager or the corporate PSM manager.

Auditors should focus on two primary tasks:

- Ensuring that the MOC procedures meet the required specifications (including the requirements established by regulations, standards, corporate guidelines, and senior management)
- Ensuring that the MOC procedures are being implemented appropriately (based on reviews of MOC records, personnel interviews, and site inspections)

MOC system audits may be part of the compliance audit requirements under the Occupational Safety and Health Administration's (OSHA's) PSM regulation, but more frequent assessments are recommended during initial implementation of the MOC system or when problems are encountered. Also, other audits or reviews of equipment, procedures, and management systems (such as comparing field installations of equipment to approved drawings) may identify the need for special MOC system audits.

Site personnel should periodically review the MOC system (by examining random samples of work that has been performed) to determine whether the correct MOC review protocol was used. They should also review process incidents and near misses that occur at the site to determine whether any MOC deficiency contributed to the incident. Appendix E provides a sample MOC audit checklist.

6.2.2 Collecting Metrics and Performing Management Reviews

Companies should consider establishing a mechanism for collecting MOC system activity data to use in populating performance and efficiency metrics. These data can be periodically reviewed by the MOC coordinator. The types of data collected depend in large measure upon the design of the MOC system documentation (RFC form). Basic information to maintain in a database includes:

- The originator's name and the date
- The process area/unit in which the change occurred
- A one-line description of the change
- Restoration date for temporary changes/repairs
- The type of change
- The name(s) of the reviewer(s)/approver(s) and the date(s)

Additional data can also be helpful in monitoring the MOC system. However, the MOC coordinator should collect only the data needed to make decisions about the system's performance and to make periodic improvements. Appendix G provides many examples of possible MOC metrics that a site should consider implementing.

Based on (1) audit results, (2) the analysis of key performance indicators, and (3) the results of management reviews, the MOC coordinator should periodically review the MOC system to determine whether any improvements should be made (e.g., coverage of new types of changes, use of more/less detailed RFC procedures, revisions to MOC system documentation). In addition, informal interviews with site personnel will often yield useful insights into how to streamline MOC procedures, fill gaps in the system, or use existing resources more efficiently.

6.3 IDENTIFYING THE NEED TO IMPROVE

6.3.1 Identifying Specific Problem Areas for Corrective Action/Redesign

As discussed above, asserting the need to redesign an MOC system does not necessarily imply the need to make the system more complex or rigorous. As outlined in Chapter 2, the goal is to maximize PSM effectiveness, which can be loosely defined as achieving the right results in a resource-efficient manner. An MOC system that is not functioning at a level sufficient to control the risks of facility activities must be repaired. However, an MOC system that is "too big of a tool" for the needs of the organization is also a concern, since this situation can divert resources from other risk control initiatives.

Therefore, the redesign of an MOC system could encompass either (1) enhancing the system or the reliability of its implementation to achieve more dependable performance or (2) responsibly trimming the complexity, rigor, or effort associated with the system to more efficiently meet the organization's needs. In reality, an MOC system is not monolithic. Different features will likely be performing at different levels, so organizations may need to both enhance and trim the MOC system when trying to fine-tune its effectiveness.

The second step (after identifying a problem) is to define the problem in sufficient detail to allow design of a solution. Table 6.4 lists the key RBPS principles and essential features for an MOC system. Most MOC performance problems are the result of either (1) failing to satisfy the intent of one or more of these key principles and essential features (or of the associated work activities listed in Appendix B) or (2) using an undue level of resources to satisfy the intent.

While some problems may be readily apparent, a gap analysis may be required to identify other areas (such as performance problems) that need to be addressed. A similar effort (an "excess analysis") might be needed to identify areas in which undue effort or the excessive use of resources is causing efficiency problems. The pertinent standards against which the MOC system is to be assessed should be unambiguously identified. For example, in an acquisition scenario, the MOC system may be performing well against the prior corporate standard, but should now be assessed against the requirements of the new owner (see Table 6.2).

When conducting a gap analysis to identify performance problems (or an excess analysis to identify efficiency problems), common sources of information would include the following:

TABLE 6.4. MOC Key Principles and Essential Features

Maintain a Dependable MOC Practice
Assurance of Consistent Implementation
Competent Personnel Involvement
MOC Practices Remain Effective
Identify Potential Change Situations
Adequate Coverage of the Scope of the MOC System
All Sources of Change Are Managed
Evaluate Possible Impacts
Appropriate Input Information to Manage Changes
Appropriate Technical Rigor for MOC Review Process
MOC Reviewers Have Appropriate Expertise and Tools
Decide Whether to Allow Change
Changes Are Authorized
Change Authorizers Consider Important Issues
Complete Follow-up Activities
Records Are Updated
Changes Are Properly Communicated to Personnel
Risk Control Measures Are Enacted
MOC Records Are Maintained

- Reports of incident or near miss investigations
- Performance or efficiency metrics
- Reports of prior, routine audits and/or management reviews
- Audits or reviews specifically commissioned for the current assessment needs, incorporating
  - interviews with personnel responsible for implementing the MOC system or those affected by the MOC system
  - reviews of MOC implementation records
  - direct observation of MOC-related activities

6.3.2 Using Performance and Efficiency Metrics

Metrics are day-to-day indicators of the health of an MOC system. These indicators are derived from, and limited by, the basic MOC system activity data that a facility chooses to collect. MOC metrics are used in a statistical process control sense; companies should experiment with collecting and using these indicators for some time in order to establish system calibration points or desired control points. Then, daily or weekly monitoring of a "dashboard" of MOC performance and efficiency metrics allows the MOC coordinator to see where imminent dysfunctions exist, which can lead to either MOC system failures that may cause accidents or system bottlenecks causing inefficiencies. The number and types of indicators that are appropriate for each site are a function of the level of detail of the MOC system and, to a degree, the level at which the MOC system is currently performing.

MOC performance measures that explicitly identify key indicators can be used to assess system performance on a near real-time basis and with more reasonable effort. Appendix F provides a substantial list of indicators that may be relevant to many MOC systems. The sensitive indicators for a specific MOC system will depend upon a variety of factors, including the MOC system design and the availability of MOC records and data. Some indicators can be used individually to help evaluate system performance, while other indicators must be used jointly.

The resources invested in operating the MOC system can be periodically reviewed, along with MOC system activity data, to determine its effectiveness. Appendix F also provides some examples of MOC efficiency metrics.

MOC efficiency indicators are derived from MOC system activity data. The number and types of efficiency metrics will be limited by the design of the MOC forms and the activity data that are collected. Appendix F also provides some examples of MOC efficiency metrics.

6.3.3 Performing Management Reviews

At facilities where the MOC review rate is very high, managers could set aside time to observe MOC reviews as they are performed. However, MOC reviews are often difficult to efficiently monitor on a day-to-day basis. Therefore, management reviews of MOC systems rely more on formal meetings with the affected parties in order to examine MOC effectiveness issues. These could be monthly or quarterly meetings. Chapter 22 of *Guidelines for Risk Based Process Safety* describes the components of a management review system. This section describes how these ideas can support effective operation of an MOC system.

In advance of the management review, the people who are responsible for a part of the MOC process should (1) conduct a self-assessment and (2) report known gaps, along with any existing plans to close those gaps. Enlisting a knowledgeable person who is independent of the MOC system (or at least a given part of the MOC process in each site area) to participate in the self-assessment is often helpful. This helps overcome one common problem with self-assessments: even the most diligent and honest person will fail to recognize a gap if he or she does not understand a requirement or recognize an issue (and, thus, this person will not be working toward closing an unidentified gap).

In addition, metrics are updated, and the MOC element owner normally makes a special effort to understand the reasons for any trends or anomalies in the metrics. Finally, if any major projects are under way to address known MOC gaps, a briefing is prepared for the management review committee.

A management review of MOC activities should demonstrate that leadership at the facility (1) is aware of and values MOC and (2) is intent on ensuring that all changes are evaluated prior to execution. The review focuses on the efforts to perform, document, collect, and maintain MOC information and metrics, including improving the efficiency of work activities supporting this element. In addition, an effective management review process educates the entire leadership team on the importance of MOC and the role it plays in helping to identify hazards, manage risk, and sustain the business.

Typical questions to ask during a management review include the following:

- Are the MOC reviews for each operating area complete and of high quality? What, if any, significant gaps were identified?
- Is the MOC system being used? Or is there evidence of circumvention?
- Did any recent audit findings address MOC? Have all corrective actions been completed or are they on schedule for completion?
- Have there been any incidents or trends for which MOC failure was a root cause or contributing factor?
- How effective is employee and contractor training on MOC? Refresher training?
- Are we using more or less staff than last year to address MOC issues?

6.4 IDENTIFYING OPPORTUNITIES FOR CORRECTIVE ACTION OR IMPROVEMENT

6.4.1 Identifying and Addressing Causal Factors

For each problem identified using the approach described in Section 6.3, management will also need to identify the causal factor(s) that created the problem. For performance problems, the challenge is analogous to that encountered during incident investigations. Performance problems will have proximate causes and underlying root causes. Identifying the proximate causes is not sufficient. The root causes of performance problems should be identified in order to properly focus remedial efforts.

For example, management review may determine that changes are frequently being implemented outside the controls of the MOC system. Investigation could reveal that this circumvention is intentional, or, alternatively that the failure to follow the MOC system is inadvertent. While this knowledge helps define the problem, further investigation would be needed to identify specific root causes that should be addressed.

Continuing this example, the following questions should be asked: What is causing personnel to intentionally circumvent the system? What can be done to stimulate compliance? The most forceful approach to solving this problem would be to implement a zero-tolerance policy that calls for disciplining anyone who intentionally bypasses the MOC system. However, such an approach ignores the possibility that one or more major obstacles are present in the MOC system that are inducing personnel to circumvent its implementation. Further investigation to identify such factors may reveal opportunities to improve the system and encourage compliance. Perhaps the MOC system is perceived to be too detailed and cumbersome. If so, is this a valid observation? Can the system be streamlined? If not, what needs to be done to help personnel understand the importance of the detailed requirements they seek to avoid?

If MOC system noncompliances are inadvertent, why is it that personnel are not using the system? Have they not been adequately trained on the requirements? Is there confusion about what constitutes a change? The more management personnel understand the true cause of the problem, the more effective they can be in developing a solution.

In reality, multiple issues to address may exist in this example. Both intentional and inadvertent noncompliances might be present, and a variety of root causes may also exist in both instances. Identifying some of the more obvious causes and stopping the investigation there might seem natural. However, management personnel should avoid the simple explanations and dig deeply enough to identify and address all the significant root causes.

Traditional root cause analysis tools, created to aid incident investigations, may be helpful when identifying the causes of MOC system implementation problems. A root cause map typically addresses issues related to procedures, training, supervision, communications, management systems, and so forth. However, the degree to which such a tool can address the cultural factors underlying performance problems may be limited. These cultural factors can be important in situations in which systemic performance problems exist. For example, the root cause "Improper performance not corrected" might be contributing to MOC noncompliances, but if such noncompliances are obvious and rampant, one should ask the question, "Why does the safety culture allow managers to fail to enforce MOC compliance?"

Once the root causes of MOC compliance problems are identified, suitable corrective actions can be proposed.

6.4.2 Identifying Typical Causes of Ineffective MOC Systems

Ineffective MOC systems are those whose performance is poor or efficiency is low. Performance refers to the degree to which MOC system implementation complies with the established requirements. Inefficient MOC systems consume too many resources in generating too few good results. Table 6.5 lists six categories of MOC effectiveness problems. While this book cannot anticipate all possible MOC effectiveness problems, this section addresses some of the more common ones associated with the categories listed in Table

6.5. Each category of problems in Table 6.5 is addressed in detail in this section (sections A through F).

A. THE SCOPE OF THE APPLICATION IS IMPROPERLY OR INADEQUATELY DEFINED

Chapter 3 described the importance of properly defining the physical and analytical scope of application when designing an MOC system. Specifically, the following three aspects were noted:

- The physical facility areas, processes, or activities to which MOC controls will be applied
- The types of changes that will be controlled using the MOC system
- The boundaries between, and intentional overlaps with, other elements or other administrative systems

The first two aspects are closely related and will be addressed jointly.

Simply put, two types of scope problems are possible: the scope was defined too narrowly or it was defined too broadly. The former poses potential regulatory and risk control problems, and the latter introduces the potential for inefficiencies in MOC system implementation.

TABLE 6.5. Categories of MOC Effectiveness Problems

- The scope of application is improperly or inadequately defined
- Implementation procedures are nonexistent, incorrect, or inappropriate in their level of detail
- Personnel are unaware of or inadequately trained on the requirements/procedures
- Sufficient resources are not available to support compliance
- Requirements are intentionally circumvented
- Problems are not identified and addressed

A.1 Scope Too Narrow

Audits against corporate or regulatory requirements may indicate that the scope was defined too narrowly. Also, safety-significant consequences resulting from incidents involving changes that fell outside the current scope may indicate a need to broaden the application of MOC controls. Such insufficiencies could be related to either the physical scope (e.g., the equipment involved) or the analytical scope (e.g., the type of change involved).

Resolving gaps in scope relative to established regulatory or corporate requirements is a straightforward exercise. More judgment and interpretation may be required to rationalize the scope in light of perceived industry best practices.

Resolving gaps in scope identified by incident histories is potentially more problematic. Most would concede that taking a continually reactive approach (i.e., incrementally expanding the scope to address the latest incident) is not acceptable. A more enlightened approach would be to respond to a significant incident (or a series of less significant events) by assessing the adequacy of the MOC system in a more general way and expanding the scope accordingly.

Such an exercise might include a review of the facility's PHAs to identify safety-critical systems (e.g., utilities or other process support systems) that have previously been excluded from the scope of the MOC system. Benchmarking against other facilities within the company, or against other companies within the industry, may also provide valuable insight.

A.2 Scope Too Broad

A variety of other factors may indicate that the scope of the MOC system has been drawn too broadly. The most common indicator may be the frequency of complaints from facility personnel that the MOC system is too complex, too time-consuming, or too difficult to implement. While such complaints might be a true indication of an overly broad scope, equally valid explanations for such opinions are discussed below. While these opinions should be listened to and considered, they should not serve as the sole determinant of excessive scope.

The scope of the MOC system should be reevaluated if persistent performance problems indicate that the tasks needed to properly implement the system cannot be reliably accomplished in time to support the required capacity (i.e., the number of MOCs that must be processed within a given time). Other factors should also be considered. Are adequate resources being provided? Are other causes of low system effectiveness plausible, such as inadequate training?

Once again, benchmarking against other company facilities or other companies may be helpful in putting the scope into perspective.

An organization with a nascent MOC system may have initially "bitten off more than it can chew." The defined scope of the MOC system may be appropriate in the long term, but may be more than the system can handle at its current level of capability. The scope may need to be carefully trimmed (in a risk-based fashion) until the organization gains experience in implementing MOC; then it can be gradually re-expanded.

A.3 Boundaries and Overlaps with Other Elements and Systems

Other possible scope-related problems stem from inadequate consideration of the boundaries and intentional overlaps that exist with other elements or administrative systems.

Chapter 2 describes the interaction required between the MOC system and other PSM elements. The imposition of inappropriate boundaries or the failure to foster the required linkages between MOC and these other elements can limit not only the effectiveness of the MOC system, but also the overall effectiveness of the PSM system. During design or redesign of the MOC system, consideration should be given to the exchange of information and the flow of work that must occur among the various PSM elements.

For example, if the need to update process safety information (in accordance with the details of a change that is processed through the MOC system) is not explicitly scoped into the MOC system design, failure to do so can result. Consequently, responsibilities may not be assigned for ensuring that this important task is addressed.

As noted in Chapters 3 and 4, other important linkages may need to be established between the MOC system and administrative systems, in addition to than the PSM system. For example, a product quality deviation might point to an analogous safety concern resulting from the purchasing department waiving feed specifications in order to obtain a lower-cost source of supply. Such an event might suggest the need for a link between the MOC system and certain functions within the procurement system.

In contrast, the MOC system should not be permitted to inappropriately overlap or infringe upon other administrative systems. For example, an organization might judge it appropriate to impose change controls on the number of graphics personnel in the design group. However, the necessity of executing a change request to authorize the relocation of a computer graphics workstation from one corner of a room to another is questionable.

Some may find it helpful to diagram the work flow for the MOC and related systems in order to identify and develop relevant linkages.

B. IMPLEMENTATION PROCEDURES ARE NONEXISTENT, INCORRECT, OR INAPPROPRIATE IN THEIR LEVEL OF DETAIL

Absent explicit regulatory requirements, written procedures for MOC system implementation may not be warranted in some circumstances; however, for the discussion presented in this section, the organization is assumed to have determined the need for written MOC implementation procedures.

B.1 Nonexistent Procedures

The absence of a required procedure can be simple to identify, but more difficult to resolve. MOC procedures should (1) address the physical and analytical scope defined for the MOC system and (2) provide sufficiently detailed instructions for identifying, classifying, evaluating, and authorizing changes. They should also address all applicable regulatory and corporate requirements.

Efforts to create (or, for that matter, revise) MOC system procedures should include seeking input from all of the groups involved (e.g., process safety organization, operations, technical, engineering, maintenance).

B.2 Incorrect Procedures

Audits or investigations may reveal incorrect instructions within MOC procedures (e.g., an interpretation or guidance that is in conflict with a regulatory or corporate requirement). Personnel who are involved in writing or revising the procedures should have a thorough understanding of applicable requirements and a good working knowledge of the mechanics of implementing an MOC system.

Some errors in MOC procedures may be based on improper assumptions or false logic underlying the requirements. A common misconception is relating the safety significance of a change (at least in part) to the estimated cost of implementing the change. History provides graphic examples of relatively inexpensive changes whose true costs stemmed from the catastrophic consequences that resulted. For example, the temporary piping jumper installed at Flixborough was a relatively inexpensive installation that resulted in catastrophic human and business costs.

B.3 Insufficient Detail in Procedures

MOC procedures should provide instructions at a level of detail commensurate with factors such as the perceived risk of activities, organizational culture, and MOC capacity. Frequent requests by the MOC coordinator for explanations of or assistance with implementing the MOC requirements may point to the need for more detailed written instructions. Another indication might be frequent errors or omissions in processing MOC requests. Determining whether the root cause of such problems is inadequacy of the procedures or insufficient training of personnel on an otherwise sound procedure is important to resolving the deficiency.

For example, many organizations find it beneficial (perhaps essential) to provide guidance within the MOC procedure on the selection of the appropriate technique for evaluating the potential health and safety effects of a proposed change. Such guidance helps determine whether a simple checklist would suffice or whether a more rigorous technique [e.g., hazard and operability analysis (HAZOP)] is warranted, based on the nature of the change.

B.4 Excessive Detail in Procedures

Excessive detail or detail that is too prescriptive can lead to inefficiencies in MOC system implementation. A large backlog of in-process MOC requests, staff complaints, or staff reluctance to comply with system requirements may be indications, but not proof, that procedures are too detailed. If this is suspected, procedure requirements should be compared to established corporate or regulatory requirements, and good practices should be determined through benchmarking. Where there is doubt, requirements should be critically challenged (by knowledgeable staff) against a common sense "Do we really need this?" criterion.

For example, MOC procedure requirements for evaluating the potential safety consequences associated with a proposed change might dictate that a HAZOP analysis be performed for all MOC requests. While a rigorous review of this nature might be appropriate for evaluating a complex process change, it would often be inappropriate for evaluating a simpler proposed change, such as a gasket material substitution.

Similarly, a procedure that is too prescriptive might mandate approval requirements that are broader than necessary for a particular type of change (i.e., the procedure requires authorizations from disciplines/departments having no relevant interest in the proposed change). In addition to potentially slowing the approval process, this situation can undermine the credibility of the MOC system.

When addressing MOC system performance problems, personnel sometimes tend to "blame the procedure" if the system does not achieve the desired results. Procedures often grow in layers, as additional instructions and details are incrementally added to address each newly discovered problem. Organizations need the discipline to resist such reactive evolution of procedures. If the procedure is truly deficient, then it needs to be addressed. However, before doing so, the organization should attempt to confirm that the performance problem is not due to inadequate training or inappropriate motivation on the part of the user.

Management might consider having two versions of a procedure: a "long" version and a "short" version. The long procedure would be more detailed, intended for the novice who is just learning the system. The short procedure would be designed for the experienced user and might take the form of a simple flowchart.

C. PERSONNEL ARE UNAWARE OF OR INADEQUATELY TRAINED ON THE REQUIREMENTS/PROCEDURES

In the absence of a conscious intent to circumvent the system, failures to process an MOC request through the system or to successfully implement MOC requirements are often attributable to lack of awareness or comprehension of the applicable requirements by otherwise well-intentioned staff.

Most personnel will have a role in MOC system implementation or will otherwise be affected by the system. At a minimum, personnel associated with the design, construction, operation, or maintenance of the facility should understand the need for – and the goals of – the MOC system, at least at a level sufficient to allow them to recognize and call attention to a potentially uncontrolled change. Other personnel will require detailed training commensurate with their level of involvement in system implementation. Initial training is required, and refresher training should be provided when warranted.

Significant or recurrent failures of the MOC system should be investigated to determine the root causes of the system breakdown. Where training issues are indicated, these should be promptly addressed. Such findings may prompt (1) a re-evaluation of the content or frequency of training or (2) the need for additional or better reference materials for system users.

The MOC coordinator often serves as a coach/tutor/mentor for MOC system users. Remedial attention should be considered if (1) this role has not been written into the job description, (2) the role has not been embraced by the MOC coordinator, or (3) the necessary resources have not been provided.

A frequent problem is the failure to identify a change as a change. MOC procedures should include clear, unambiguous definitions of both change and replacement-in-kind. Training and/or reference materials should provide lists of facility-specific examples of each to make them more relevant to the user.

Another common source of problems is the process for assessing the potential safety impact of the proposed change. For complex and/or potentially significant changes, this assessment may require a full-fledged PHA. Or, at a minimum, those personnel who are designated as MOC reviewers may require hazard evaluation training similar to that provided to PHA team members/ leaders.

D. SUFFICIENT RESOURCES ARE NOT AVAILABLE TO SUPPORT COMPLIANCE

Problems commonly related to resource limitations include the inability to develop and process MOC requests quickly enough to meet the need of the program or to keep pace with the associated activities. Examples include completing tasks that are prerequisites to starting up modified facilities (such as training affected personnel) or performing follow-up activities such as updating process safety information (PSI). The inability to keep up with the demand for MOC requests might tempt personnel to circumvent the system. The inability to complete associated activities in a timely manner risks compromising the integrity of other RBPS elements.

MOC system implementation can be resource intensive, with the required system throughput/capacity having a direct impact on resource requirements. Large facilities (and smaller facilities experiencing more frequent changes) may find it necessary to augment staffing and, perhaps, acquire new technology to support the information/document work flow and management tasks. Common resource issues are discussed below; the use of newer technologies is discussed in Section 5.6.

In addition to the need for trained personnel to initiate, develop, review, approve, and implement change requests (see Section 5.3), personnel will be needed to support other MOC system activities. Such personnel include, but are not limited to, draftsmen to update piping and instrumentation drawings (P&IDs) and other design documents, procedure writers to update operating and emergency procedures, and technical staff to perform engineering evaluations.

Required resources also include the management systems needed to implement the function. For example, providing updated P&IDs for an MOC request is made more difficult if a systemic problem exists with P&IDs already being out of date. Similar concerns can exist about other PSI, operating procedures, and so forth. The work products associated with these other PSM elements may require significant remedial attention in order to support improvements in the implementation of the MOC system.

Performance problems, such as an excessive backlog of P&IDs requiring updating, should be investigated to identify their root causes. For example:

- Does the problem stem from resource limitations?
- Is it an issue of failing to establish proper priorities?
- Does the situation reflect insufficient coordination between two RBPS elements or functions?

E. REQUIREMENTS ARE INTENTIONALLY CIRCUMVENTED

As previously noted, failing to process a change through the MOC system or omitting a particular procedure requirement for a given change can be inadvertent, due to lack of awareness or comprehension of the requirements. However, such failures or omissions can also be intentional. The circumstances leading to widespread, intentional circumvention of an MOC system can be among the most difficult to address. To do so requires determining the factors that are behind this intentional violation of established requirements.

At the core, such intentional violations reflect an inadequacy in the organization's safety culture. An extreme case would be one in which the disregard for MOC requirements is widespread, readily apparent, and likely sanctioned (at least tacitly) by facility management. The safety culture of an organization is significantly influenced by the expressed beliefs, shared attitudes, and demonstrated actions of management. A profound lack of safety leadership can cascade down through the organization in the form of general indifference to established safety requirements.

Proposing specific solutions to such situations is beyond the scope of this book. Generally, however, the solution should come from top management in the form of strong support for safety issues and programs and intolerance for intentional safety violations. In extreme situations, absent a new self-awareness of responsibilities on the part of facility management, the stimulus for correction may have to originate from outside the organization.

Less egregious circumstances are more common. Other failures to implement MOC requirements, while intentional, may be caused by situational drivers (real or perceived) that prompt someone to believe that the justifiable (or, perhaps, only) alternative is to circumvent the system. This is not meant to justify the person's actions/inactions. However, understanding the motivation behind such action is necessary in order to address its causes.

Experience has shown that rule-driven, zero-tolerance approaches to addressing such behavioral problems are only briefly effective, typically when the likelihood of being detected by an enforcing authority is thought to be high. Behavioral change is most effective in conjunction with cultural change. Removing impediments (real or perceived) to acceptable behavior can be a catalyst for such changes.

The importance of safety culture to the successful implementation of PSM systems, including the MOC element, is addressed in the Center for Chemical Process Safety book *Guidelines for Risk Based Process Safety*.

E.1 Value of MOC Not Appreciated

MOC is labor intensive and, often, facility staffing is decreasing as workload is increasing. Introducing a new MOC program, or re-engineering a program that will result in additional responsibilities and time demands, runs the risk of being regarded as the straw that broke the camel's back. Consequently, audits, interviews, or investigations often reveal a lack of buy-in; that is, reluctance on the part of some personnel to fully engage in the implementation of the MOC system.

Personnel should understand the importance and benefits of MOC implementation so that a value for successful MOC implementation becomes part of the culture. Successful cultural change requires that expectations of new attitudes and behaviors be communicated and reinforced, that these new attitudes and behaviors demonstrate successful results, and that the members of the organization recognize and appreciate the resulting successes. The benefits of a sound MOC system were addressed in Chapter 1. Management should be sensitive to employee attitudes and step in as advocates for the MOC system when needed to stimulate employee involvement.

E.2 Requirements Too Complex or Too Broad

This issue, closely related to the preceding one, is often the basis for a passive-aggressive resistance to complying with MOC requirements: "I'm all in favor of MOC, but you've made it too difficult . . . the procedure is too long . . . I'm not buying it."

Some employees might feel that the MOC system contains too many tedious details. Assertions that the system is bigger than it needs to be should be carefully evaluated, with the goals of reducing staff resistance and increasing system efficiency. In the end, however, the system must be what it must be. Managers may need to stress the benefits of, and rationale for, the attributes of the system that prompt the greatest concern.

E.3 Perceived or Actual Resource Limitations

Except under certain heroic circumstances, human nature is to not attempt the impossible and to avoid, if possible, the really difficult. The perception, justified or not, that insufficient resources have been provided for successful implementation of the MOC system can discourage employee involvement.

To maintain MOC system credibility, management should not assign responsibilities without providing the resources to satisfy them. For example, holding the MOC coordinator responsible for ensuring that P&IDs are updated, but not providing enough draftsmen to do so, creates the risk that an increasing number of MOC request forms will indicate that P&ID updates are not required.

Time is everyone's most important resource. If MOC system implementation requires the assignment of additional substantive responsibilities, this may prompt anticipation that other responsibilities will be reassigned. If this is not the case, management should clearly establish priorities among the new MOC system responsibilities and existing responsibilities.

Resource shortfalls may be actual or imagined. Assertions about shortfalls should be carefully evaluated to determine the facts of the matter. If resources are indeed limited, management may need to either address the shortages or reevaluate the scope of the MOC system. Before doing so, however, management should try to determine whether existing resources are being used as efficiently as possible (e.g., Would additional staff training result in more efficient use of existing resources?).

Increasingly, new technologies are also being applied to improve the efficiency of MOC systems.

F. PROBLEMS ARE NOT IDENTIFIED AND ADDRESSED

While the logic may sound circular, one of the most frequent sources of problems in MOC systems is the failure to identify and address problems in the MOC system. Methodical audits and thorough incident investigations may reveal MOC system problems, but audits are typically infrequent and major incidents hopefully occur even less frequently. MOC systems, because of their complexity and importance within the PSM system, need to be monitored more frequently. This is particularly true in the case of systems in transition (i.e., new systems being implemented, or re-engineered systems implementing corrective actions). Management reviews of appropriate frequency should be implemented to provide a more real-time, ongoing assessment of the health of the MOC system. Chapter 7 provides more perspective on this issue.

Another good practice for some organizations is regarding failures to process changes through the MOC system as near misses that require investigation to determine the root causes of the failure.

6.4.3 Performing a Gap Analysis for Performance Issues

Assemble a team to contrast the design specification established in Chapter 3 with the description of current practices assembled in Section 5.3, noting any gaps. Not all practices will be documented in the written procedures. If the review team is confident that an established, undocumented practice exists, this should be noted (consider documenting the practice in the MOC procedures, if it is substantive). In addition, all documented procedure requirements may not actually be routinely implemented. If the team is aware of such situations, these too should be documented for resolution.

6.5 IMPLEMENTING THE REDESIGN/IMPROVEMENT EFFORT

If company personnel want to improve MOC, they need to be motivated, have access to adequate information, decide on an execution approach, and then make it happen. Effective MOC improvement requires the following elements:

- An identified need
- A management plan
- Allocation of resources
- Measurement to confirm improvement or implementation

In some cases, MOC system improvements will require repetition of some of the steps presented in Chapters 3 through 5.

6.5.1 Managing the Redesign Effort

The following paragraphs suggest one approach to managing the MOC system redesign effort. All of these steps may not be applicable to a given situation, and other tasks not identified here may be appropriate to meet specific needs. A preliminary redesign team should be formed and charged with making the necessary improvements in system implementation.

Establish or reaffirm the goals/intent of the MOC system. This initial and essential step provides the foundation for the effort by focusing the preliminary redesign team on the goals of the MOC system. The team should identify relevant sources of performance requirements, such as (1) applicable regulations, (2) corporate requirements, and (3) facility-identified objectives. Most likely, the first of these items will be nonnegotiable, as may the second. However, facility-identified objectives may be subject to reevaluation and certainly should be confirmed as being consistent with the first two items.

If a design specification (as described in Chapter 3) for the MOC system was prepared originally and still exists, it should be updated in this and subsequent steps. If no such specification exists, the team may want to consider generating one as part of the redesign effort.

This step should produce a team consensus of what successful MOC system implementation should look like and the results it should produce.

Gather relevant information regarding MOC system performance. Presumably, some information about less-than-desirable MOC system performance has prompted the redesign effort. This information should be supplemented by other relevant, available information. This might include

- Audit reports
- Documentation of management reviews
- Incident investigation reports
- Employee suggestions
- Safety committee meeting minutes
- Trended performance metrics
- Records of staff training on MOC system implementation (i.e., training on the program itself, not training on any particular change)

Gather all MOC procedures, forms, training materials, and reference documents. Make certain that you have all of the relevant system information, procedures, and data.

Supplement the redesign team, as appropriate. A review of the performance data compiled in Section 6.2 may identify areas of inquiry that require input from particular groups or functions not represented on the preliminary redesign team. Emphasis should be placed on staffing the team with the right mix of expertise and experience related to the MOC system. Those most experienced with the system may have the greatest biases or vested interests, so emphasis should be placed on selecting open-minded individuals. Management might consider selecting someone to lead the team who is independent of the MOC function, and perhaps, of the facility.

6.5.2 Itemizing and Evaluating Known Concerns

MOC failures, issues, or problems identified can be tabulated using the guidance in Sections 6.3.1 through 6.3.3. Prioritize these issues based on their perceived significance and, where feasible, identify the immediate and root causes of these concerns.

Appendix G lists some of the more common causes of MOC performance problems. Traditional root cause identification techniques (e.g., the Five Whys technique) may also be helpful. As with incident investigations, the goal is to identify and address the underlying systemic causes (root causes) of performance problems.

6.5.3 Proposing Corrective Actions to Address the Causal Factors

Corrective actions should be developed to address each of the causal factors identified in Section 6.4.1. Some general classes of corrective actions include:

- Adjusting the scope of application for the MOC system
- Providing new or revised procedures
- Providing new, modified, or more frequent training
- Providing new or revised reference materials (e.g., work flow diagrams charting the processing of a typical change)
- Rationalizing the resources available for new system requirements
- Providing additional management oversight

Examples of MOC system corrective actions are provided in Appendix G, and a diagnostic tool is provided in the form of an Excel spreadsheet.

6.5.4 Repeating the Evaluation to Address Efficiency Issues

The team should next address efficiency issues associated with the MOC system by substituting an "excess analysis" for the gap analysis described in Section 6.4.3, then repeating the steps in Sections 6.4.1 and 6.4.2.

6.5.5 Challenging the Proposed Revisions to the MOC System

The team should test the proposed modifications to the MOC system. A step-by-step review of the revised procedures – determining whether each requirement is necessary and sufficient to achieve its intended objective relative to the design basis for the MOC system – should provide additional confidence in the proposed redesign. This review can be done informally, or in a more structured fashion by analyzing a detailed MOC workflow diagram using a technique such as "procedural HAZOP" or "six sigma FMEA".

The test for necessity is not meant to imply that the redesigned system should strive for mere compliance with the minimum established MOC system requirements. Many valid reasons may be provided for redesigning the system to achieve performance beyond these minimum requirements. However, management should recognize that each such increment adds to the complexity of the overall system and could divert limited resources from other RBPS initiatives. Justification should go beyond "it sounds like a good idea" to "it is a prudent investment of resources."

6.5.6 Implementing and Monitoring the Redesigned or Improved MOC System

When implementing the revisions to the MOC system, management should assess the scope and breadth of the changes and how they should be rolled out. Is it practical or advisable to implement all the changes at once? Or should the changes be prioritized and implemented over time? If a phased implementation is selected, which revisions should be packaged or sequenced for rollout? What training will be required to familiarize personnel with the new system?

Leading and lagging performance indicators appropriate to the new system should be implemented and tracked (see *Guidelines for Risk Based Process Safety*). Metrics should be tracked and trended more frequently during the early stages of implementation. As confidence grows in the performance of the redesigned system, the monitoring frequency can be decreased.

User/customer feedback should be encouraged during implementation. Providing periodic management reports may be appropriate, addressing:

- The implementation status
- Significant issues/variances encountered
- Remedial actions planned or completed
- Implementation milestones and success stories

7

THE FUTURE OF CHANGE MANAGEMENT

Process safety management (PSM) has matured during the past 15 years. Yet many opportunities still exist for continuous improvement in design and operation that promise even more effective management systems in the future. Management of change (MOC) remains one of the most important PSM elements – and one of the most difficult to implement and keep healthy. Many companies operate MOC systems for a variety of non-process safety reasons. The authors hope that this book will stimulate management's thinking about effective ways to improve an organization's MOC activities.

Experience is a powerful teacher; yet the painful lessons learned from watershed events are all too quickly forgotten. Recent learnings have shown that gaps can occur in MOC system implementation because the pressures of everyday business can overwhelm the lessons of history. Future MOC systems should be designed to be more fault tolerant and to have effective, built-in redundancy. Facilities should adopt practices that nurture a safety culture. Management should maintain process safety competency and resolve the "loss of corporate memory" prevalent in industry, which can hamper proper MOC system operation.

Supervision and workforces alike should embrace operating discipline as an essential feature of improving and securing human performance. Metrics should be used to realize the highest return on every process safety resource invested in MOC, and management should commit to periodically reviewing the MOC system in order to make continuous improvement real, not just a slogan.

Layered, effective MOC system control functions (using metrics, management review, and audits) should be viewed as management tools for organizational learning, as originally intended, not for placing blame.

The business case for process safety and MOC system implementation should be established so that safety, health, and environmental issues can be managed in the same manner as sales, raw materials, inventories, and capital. MOC practices should pervade company operations throughout the life cycle of equipment, processes, and sites. MOC practices should be adopted by ALL industries that manufacture or use hazardous chemicals or energy, and their use should become standard practice.

Recognizing that "good things happen through planning, while bad things happen all by themselves," the process safety community should apply one of its strongest diagnostic tools – root cause analysis – to its broken or underperforming MOC processes, procedures, and practices. Just as a root cause analysis related to an incident investigation seeks to identify specific management system root causes, an MOC root cause analysis should look for system-wide management issues. In addition to supporting incident investigations, root cause analysis should be used for evaluating undesired MOC outcomes and addressing MOC performance and efficiency problems.

MOC tools need to improve so that non-experts can competently use them. Companies and facilities should develop expert systems to assist with real-time MOC risk decisions. The Center for Chemical Process Safety's PSM Web community should grow, affording seamless virtual connectivity between workforces, facilities, companies, industries, and countries. This would allow everyone to benefit from lessons learned and to benchmark MOC practices in real time. MOC systems should be fully electronic, with work flow tools for communicating, controlling, and managing MOC effectiveness (even integrated with distributed process control networks).

Table 7.1 lists some areas in which change management may evolve during the next decade.

TABLE 7.1. Possible MOC Growth Areas

- Totally electronic MOC systems
- MOC systems that dovetail so perfectly with hazard/risk studies that MOC reviews will automatically update the current version of the site PHA or risk study
- Expert system tools to aid MOC reviewers in evaluating the risk of a proposed change
- Management of newly recognized important sources of change (e.g., culture or organization)
- MOC systems that are operated "virtually" from distant corporate locations where the necessary MOC hazard review resources are available
- MOC systems at various geographical locations that are interconnected to allow sharing of MOC experiences and harmonizing of changes in similar site processes
- MOC systems that communicate among different companies in order to share new hazard/risk information and lessons learned
- MOC systems that dovetail with regulatory compliance submittal software that will automatically update resubmissions (e.g., risk management plans)
- MOC systems that address changes triggered by outside sources
- Fully integrated PSM work flow systems, including MOC
- Expansion in the technical areas in which MOC is implemented, and integration of all MOC systems (e.g., process safety quality, environmental, security
- More prevalent self-auditing and management reviews
- Integration with work order and purchasing systems
- More interaction with suppliers and customers
- Measures for tracking continuous improvement of MOC

APPENDIX A:

EXAMPLES OF REPLACEMENTS-IN-KIND AND CHANGES FOR VARIOUS CLASSES OF CHANGE

The table in this appendix provides examples of replacements-in-kind (RIKs) and changes for various classes of items that undergo changes. The categories below are listed in order of expected frequency of use:

- Process equipment
- Process controls
- Operations and technology
- Procedures
- Safety systems
- Maintenance and inspection requirements
- Site infrastructure
- Organization
- Policies
- Other process safety elements

Not all changes are necessarily less safe than the original design. Also, some companies have management systems other than MOC systems that control some of these classes of changes (e.g., staffing or procedural changes). In those cases, the alternate systems need to satisfy the MOC system requirements for the specific class of change. Consider using these examples, or generating your own context-specific examples, when developing MOC awareness-level training for site personnel.

Replacement-in-Kind	Change	Example Concern with Change*
Process Equipment		
Replacing vessels or piping with equipment having the same dimensions, configuration, metallurgy, wall thickness, pressure and vacuum rating, design temperature, heat treatment, etc.	Changing from carbon steel to stainless steel	Potential for pitting in chloride-bearing service
	Changing from Schedule 40 to Schedule 80 piping	Piping flexibility analysis may be invalidated by heavier gage piping
	Changing pipe diameter	Higher flow resistance/erosion for smaller diameter piping. Lower flow rate in larger piping may result in solids falling out of suspension in slurry flow
	Relocating the vessel vent nozzle	May result in inadequate inerting of head space in the purged vessel
	Replacing the reactor with one of equal volume but different length-to-diameter ratio	Changes in vessel mixing and heat transfer characteristics
	Changing from ANSI Class 900 flanges to ANSI Class 1500 flanges	Putting additional stress on other parts of the piping system due to heavier flange
Repairing a corroded vessel to restore its original wall thickness	Derating the vessel to operate at a lower pressure consistent with the decreased wall thickness	Need to rerate the relief valve, with a consequent decrease in relieving capacity
Replacing a valve with one that is, in *all* respects, identical to the original – OR – Replacing a valve with one that meets all of the design specifications for the original valve	Replacing a rising-stem gate valve with a quarter-turn ball valve	Invalidates the procedure calling for the operator to throttle flow by rotating the valve handle X number of turns
	Other changes in the style of the valve (e.g., replacing a globe valve with a butterfly valve, changing trim characteristics)	Potential changes in the tightness of shut-off, tendency to plug, maintenance requirements, flow/pressure drop characteristics, etc.
Replacing rotating equipment with new equipment of the same material, capacity, flange rating, seal design, driver type, horsepower, etc.	Changing material of construction including internal parts	Corrosion/erosion
	Increasing the impeller size	Downstream equipment may not accommodate potentially increased flows. Increased head lifts downstream PSVs
	Using a single seal to replace a tandem seal in a pump	Different spare parts and maintenance requirements. Greater potential for seal leak

Replacement-in-Kind	Change	Example Concern with Change*
	Increasing driver horsepower	Increased motor electrical requirements
	Changing from an electric motor to a steam turbine driver (or vice versa)	Utility failure response scenarios should be factored into emergency procedures
	Temporarily replacing a centrifugal pump with a positive displacement pump while the original pump is out of service	Need for a reliable relief path for downstream piping
Making a piping system repair in a fashion that conforms exactly to the original design specification	Replacing a section of pipe with a hose	Potential for a lower pressure rating for the hose. Greater potential for physical damage. Hoses in high pressure service whipping about if they become disconnected
	Adding a valve where there was none	Potential to invalidate operating procedures. New cause for low/no flow. New leak/release point (especially for vents and drains). Potential to block in a piping section, which would then require overpressure protection
	Repairing a service or process leak by banding or via an engineered clamp requiring the injection of a sealing compound	Need to confirm that the pressure rating for the temporary repair is adequate for the service. Potential incompatibilities between sealant and process materials
	Substituting plastic pipe for steel pipe	Potential for static electric charge accumulation and potential ignition of flammable vapors or combustible dusts
	Replacing a glass fiber reinforced PTFE gasket with an unreinforced PTFE gasket	Potential for cold flow of the gasket, leading to flange leaks
	Replacing a spiral-wound gasket with a conventional sheet gasket in flammable liquid service	Sheet gasket is more likely to fail in a fire-exposure situation, potentially worsening the event
	Replacing steam heat tracing with electrical heat tracing	Introduces new failure modes upon loss of utilities. May need to provide an alternate means of overtemperature protection

Replacement-in-Kind	Change	Example Concern with Change*
	Adding a new branch header to a facility dust collection/removal system	Potential to reduce linear velocity through the system to the point that solids collect in the system
	Replacing an electric motor with one having a different electrical classification or running temporary power wiring to a temporary pump	New installation may not satisfy area electrical classification requirements
	Temporarily replacing the permanent piping insulation with glass fiber batting	Susceptibility of the replacement insulation to absorb water, leading to corrosion under the insulation
	Replacing vessel insulation with non-fireproof insulation, or replacing the fireproof insulation cover with a non-fireproof insulation cover	May affect the PSV design basis. May decrease the time to maximum rate during a fire scenario for a reactive chemical
Process Controls		
Resetting the trip point on an interlock, but staying within the safe operating range established by prior safety analyses	Resetting the trip point beyond the safe operating range	Potential to establish unsafe process conditions
	Adding a new alarm (e.g., high temperature deviation alarm on a chemical reactor)	Need to train operating personnel on the proper response to the new alarm. Potential for exacerbating a situation where process alarm overload already exists
	Bypassing an interlock using the DCS system	Loss of interlock protection. Interlock bypasses are often easier to implement within a DCS, and personnel may be more prone to do so

Replacement-in-Kind	Change	Example Concern with Change*
Tuning a controller to more tightly control the process variable	Installing additional field hardware in a conventional pneumatic control system to tie several control loops together in a cascaded control scheme	Need to update operating procedures and train operating personnel. Need to update P&IDs
	Modifying the programming within the DCS system to tie several control loops together in a cascaded control scheme	Functionally equivalent to the case above. In addition, there will be a need to update the documentation of the DCS programming
Replacing a DCS system component with an identical replacement (e.g., a module that receives field inputs and converts the analog signal to a digital signal for use within the DCS)	Replacing the same module with an upgraded version with enhanced diagnostic capabilities or one that provides new options for responding to bad input signals	Prior assessments of DCS failure modes may have been based on the functional description of the original module and may have to be updated
	Relocating a field signal from one input module to another	Relocating the input may defeat the redundancy provided in the original DCS configuration (e.g., the move may combine both level signals for a vessel on the same input module, potentially resulting in the loss of both signals if the common input module fails)
Replacing an instrument with an identical spare	Replacing a vortex shedding flowmeter with a magnetic flowmeter	Magnetic flowmeter may be unable to withstand the service conditions
	Replacing a transmitter that produces an analog output with one that produces a digital output. The transmitter is associated with an interlock	May require the development of new maintenance procedures, if such a transmitter has not been previously used on site. Change may have to be reviewed to determine the potential impact on the reliability of the interlock
	Reducing the range (span) on a transmitter calibration. New range is well within the safe operating limits established by prior safety analyses	Recalibrated transmitter may not properly cover the entire range of conditions that could be experienced within the process (e.g., the process may experience a temperature beyond the upper limit of the calibration, leaving the operator unaware of the state of the process)
	Moving a conventional analog control loop into the DCS	Requires procedure revisions and retraining of operating personnel

Replacement-in-Kind	Change	Example Concern with Change*
Operations and Technology		
Modifying process operating parameters but staying within the safe operating range established by prior safety analyses; including, but not limited to: • Flow • Temperature • Pressure • Composition • Time (mixing, settling, reaction) • pH • Speed/RPM • Production rate • Inventory • Weight • Level • Density • Frequency/amplitude of vibrating equipment • Voltage/current/power	Increasing process throughput beyond the currently established unit nameplate capacity	Potential impact on relief system capacity requirements. Higher operating pressures or inventories may invalidate the basis for prior facility siting studies
	Increasing velocity in process lines beyond the established limit	Potential for increased erosion or electrostatic charge generation
	Increasing the temperature of a tank cleaning step above the established limit	Potential for exceeding the decomposition temperature of residual material in the vessel. Potential for exceeding the flash point of the cleaning solvent
	Initiating a processing step above atmospheric pressure instead of under vacuum to increase production rates	Higher pressure causes higher temperature, potentially resulting in exceeding the maximum safe operating temperature for the reaction. Higher potential release rate if a leak occurs
	Raising a tank level above the established safe high level limit to temporarily accommodate more inventory	Potentially exceeding the regulatory-based inventory limits. Potential to exceed maximum service pressure in lower portions of the tank (established based on corrosion monitoring of tank shell)
	Reducing the concentration of a polymerization inhibitor in a monomer product	Potential for runaway polymerization while in storage or during transport
	Reducing the reactor settling or phase separation time below the established limit	Loss of product or excessive organic carryover into aqueous waste streams, increasing the burden on the waste treatment facility
	Changing machine settings on rollers, crimpers, or extruders beyond the established limits	Increased heat from friction. Premature bearing failure. Increased fine particle generation

Replacement-in-Kind	Change	Example Concern with Change*
Adhering to process flow paths previously assessed by PHAs or other safety evaluations	Temporarily operate bypassing a process equipment item such as: • Heat exchanger • Separator • Knockout pot	Potential for equipment damage due to: • Low temperature embrittlement of downstream equipment (lack of heating) or exceeding the maximum design temperature of the equipment (lack of cooling) • Loss of liquid seal, allowing high pressure gases to reach downstream vessels having lower design pressures • Carryover of liquid into a compressor
	Adding a new flow path for a high-pressure flammable chemical without providing adequate isolation	Potential for uncontrolled releases of the flammable chemical in the event of a loss of containment incident
	Using any piping jumper (process or service) for a production change	Redirected flows may introduce potential process upsets that have not been previously assessed, and for which engineered or administrative controls may not have been provided
	Adding a new injection quill to introduce a treatment chemical into the process (e.g., a corrosion inhibitor)	Potential incompatibilities between the treatment chemical and this, or a subsequent, process stream. Potential for accelerated injection point corrosion, leading to loss of containment

Replacement-in-Kind	Change	Example Concern with Change*
Modifying the packaging of raw materials, intermediates, or products, <u>where the new packaging satisfies the requirements established for the safe handling of the subject material</u> (e.g., substituting a plastic drum for a stainless steel drum for a corrosive solution when the only requirement is for the use of a corrosion-resistant container)	Receiving an organic powder in 1-ton plastic flexible intermediate bulk containers (FIBCs) instead of the 50-pound paper bags used previously	Mechanical hazards of handling and unloading larger, heavier containers. Need to establish controls on the rate of unloading of the FIBCs to prevent excessive electrostatic charge generation and potential ignition of flammable vapors present
	Packaging (or repackaging) a solution of an organic peroxide polymerization initiator into a larger container	Larger container is less effective in dissipating heat from decomposition of this thermally unstable material. Temperature rise in the container may exceed the self-accelerating decomposition temperature of the peroxide solution
	Temporary receipt of a highly toxic material via tank truck rather than the railcars normally used	Potential for higher transportation risks using over-the-road trucks rather than railcars. Smaller transport vehicles result in more connections and disconnections of unloading lines, increasing the potential for errors resulting in releases. Engineered protections for railcars (e.g., car motion detectors interlocked to isolation valves in unloading lines) may not be available for tank truck unloading
Using alternative vendors as sources of a feed stock that meets all established purchase specifications	Changing to a more reactive catalyst type recommended by the traditional vendor	Higher reaction rate may exceed the cooling capacity of the reactor, potentially leading to a runaway reaction
	Purchasing feed stock outside of established purchase specifications as a lower-cost alternative to the normal supply	Potential for safety-related processing problems, such as higher corrosion rates leading to loss of containment or possible side reactions
Rearranging warehouse stock, but within the established basis for safe operation with respect to considerations such as: • Inventory limits	Exceeding the maximum volume permitted for bulk storage of a thermally unstable solid	Larger pile is less effective in dissipating heat from decomposition of this thermally unstable material. Temperature rise in the pile may exceed the self-accelerating decomposition temperature for the solid

Replacement-in-Kind	Change	Example Concern with Change*
• Compatibility groupings • Fire protection system capabilities	Temporarily storing drums of a water-reactive material on the warehouse loading dock	Inadequate protection of containers and contents against exposure to rainwater
	Purchasing and storing larger-than-normal quantities of drummed flammable liquids in order to take advantage of a quantity discount	May exceed the flammable loading assumed for the design of the warehouse fire protection system
	Storing a new chemical without evaluating its hazardous properties	Potential for storing the new chemical beside one or more incompatible materials
Procedures		
Making minor editorial changes or typographical corrections to operating or maintenance procedures	Modifying operating procedures to reduce or eliminate operator rounds in an area	Loss of the benefits of operator presence, such as leak detection and alertness to other nonstandard conditions
	Changing phone numbers on an emergency call-out list	Need to notify affected personnel of the changes. Need to ensure that all emergency response manuals are updated to include the revised list
	Changing previously established safety, quality, or operating limits in the operating procedure	Need to train affected personnel on the revised limits. Need to revise other interrelated PSI, such as separate tables documenting safe operating limits consequences of deviation, or steps to avoid deviations. See **Operations and Technology**
	Relocating information on safe upper and lower limits, consequences of deviation, etc., from the operating procedure to a separate referenced document	Need to notify affected personnel of the changes. Need to modify management systems to ensure that the referenced documents and the operating procedures are periodically confirmed to be accurate, up to date, and consistent in content
	Making substantive formatting changes in the facility operating procedures	Need to assess the new format for human factors considerations (e.g., would the new format decrease or increase the potential for misinterpretation and operator error?)

Replacement-in-Kind	Change	Example Concern with Change*
	Moving from a hard-copy based operating procedure system to an intranet based one	Need to train all affected personnel on how to access/use the new system. Need to establish contingency plans for accessing procedures in the event of an intranet outage
Sampling a process stream on Mondays and Thursdays instead of Tuesdays and Fridays (assuming that other related activities are constant throughout the week)	Sampling a process stream once a week instead of twice a week	Potential to not detect process deviations of shorter duration
Using OEM manual maintenance procedures	Using site-generated maintenance procedures instead of OEM manual maintenance procedures	Need to confirm that the site procedures address all the appropriate and necessary maintenance tasks for the equipment item involved. Need to train all affected maintenance personnel on the new procedures
Delegating authorization responsibilities (e.g., for work order approval) to a properly qualified substitute in accordance with the pre-established delegation schedule	Modifying an existing authority delegation schedule to assign authorities to a lower level in the organization	Ensure proper training for personnel commensurate with their new responsibilities
	Changing the purchase order approval practice to delete the requirement that a particular department be involved in the approval	Ensure that the management system still involves appropriate disciplines and areas of expertise in the approval process
Safety Systems		
Scheduling process outages as required to provide access to safety systems for ITPM	Adding an isolation valve beneath a pressure relief valve to make it easier to remove and test the relief valve between process outages	Provides a new mechanism for defeating the protection intended to be provided by the pressure relief valve. May require revising the operating procedure and updating the chain lock or car seal list
Recharging a fixed fire protection system with the same firefighting agent previously used	Replacing a Halon® system with a CO_2 system	Should confirm that the new system is sufficient to address the types and quantities of combustibles and flammables present. Should provide new maintenance procedures and training for maintenance personnel
Replacing an explosion relief vent panel with an identical unit from the same manufacturer	Replacing an explosion relief vent panel with a panel having a higher burst pressure to "prevent spurious openings"	Potential loss of intended protection and consequent damage to the "protected" vessel

Replacement-in-Kind	Change	Example Concern with Change*
	Installing new equipment within the discharge path of an explosion relief vent panel	Potential exposure of personnel to hazardous conditions if there is a discharge from the vent. Potential to degrade the performance of the vent
Replacing a relief valve with a new valve that is either identical to the original or meets all the relevant design and performance specifications established for the valve	Redirecting atmospheric relief valve discharges to an existing flare header	Relief path may not have adequate flow capacity due to the back pressure in the flare header. Potential to exceed the design flow capacity of the header/flare. Potential to raise the pressure in the flare header sufficiently to significantly reduce the effective capacity of the relief valves discharging to it (e.g., in a unit fire scenario)
	Replacing a balanced bellows relief valve with a conventional relief valve	Loss of backpressure compensation on the relief valve and reduced discharge capacity. Increased potential for the relief valve to open prematurely if operating close to its set pressure
	Reducing the opening pressure of a relief valve in vapor service in an attempt to better protect the vessel it serves	Reduced flow capacity for the valve at the lower operating pressure may not be adequate to protect the vessel
Operating a process with an interlock out for maintenance but with alternative means of protection provided, as specified in the operating procedures	Continuing operation with an essential safety system (e.g., a relief valve) out of service but with no alternative means of protection provided	Loss of the required protection. Implication to staff that operating with safety systems defeated is acceptable
Inspection, Test, and Preventive Maintenance Requirements		
Changing from a spring turnaround to a fall turnaround within the run-time limit for the unit	Postponing a unit turnaround beyond the design run-time limit	May result in exceeding maximum allowable intervals for certain equipment tests and inspections, or may push equipment (e.g., thin piping) beyond the estimated remaining service life
Changing to another brand of lubricant that meets the specifications established for the particular service required	Changing to a lubricant that is outside current specifications	Suitability of the lubricant for the intended service conditions. Potential need to revise lubrication frequencies

Replacement-in-Kind	Change	Example Concern with Change*
Decreasing the test interval for a high temperature interlock	Increasing the test interval of a high temperature interlock	Increasing the test interval increases the probability that the interlock will fail undetected between tests. (Note: Testing the interlock requires taking it out of service during the test and introduces some potential for human error in conducting the test. Excessively short intervals may actually lessen the reliability of the protective system. Accordingly, some organizations might regard any revision of a test interval to be a change.)
Reassigning ITPM tasks to comparably qualified personnel within the same work group	Installing online compressor vibration monitors (transmitting readings to the control room DCS) and discontinuing the current practice of direct readings taken by the rotating equipment maintenance group	Need to train operators to be aware of the significance of the new readings on the DCS system and their proper response to nonstandard conditions. Need to establish new systems for making the necessary information available to those responsible for detailed evaluations of compressor performance
	Reassigning certain maintenance tasks from maintenance personnel to operators	Need to provide operators with the appropriate procedures, tools, and training for their new responsibilities
	Discontinuing certain ITPM tasks	Potential for reducing the reliability of the subject systems
Increasing inspection frequency based on accepted engineering methods (e.g., remaining life calculations) in accordance with established facility procedures	Reducing maintenance frequency based on resource constraints without considering past operating experience	Potential for increasing the likelihood of unanticipated equipment failures
	Changing the inspection method for unit piping thickness from ultrasound to X-ray	Need to establish the basis for confidence that the alternative method achieves the required objectives. Introduction (or more frequent possibility) of a new hazard (i.e., ionizing radiation) and the need to train personnel on the hazard

Replacement-in-Kind	Change	Example Concern with Change*
Site Infrastructure		
Replacing process area lighting with fixtures of the same type and design	Installing a new light or installing a new type of lighting fixture in an electrically classified area	New/revised installation may violate electrical classification increasing the potential for a fire or explosion in the event of a flammable release
	Relocating lighting in a process area to better illuminate process equipment	Changes in lighting patterns may invalidate some portion of the security vulnerability analysis. May impact emergency response/evacuation issues
Providing a new building or relocating personnel within existing buildings when the buildings are beyond the consequence zone for process indents (vapor cloud explosions, toxic releases, etc.)	Adding a maintenance planner's office next to the process control room for monitoring contract maintenance activities	Introduces a new source of personnel risk not addressed in prior facility siting studies
	Increasing the occupancy of the site control room building	Increases the risk associated with building occupancy beyond that assessed as tolerable by prior facility siting studies. Increased population may exceed the building systems' capability to protect occupants during an emergency (e.g., the capacity of emergency breathing air supply required during a shelter-in-place scenario)
	Modifying existing buildings (e.g., structural modifications, adding a heavier air conditioning unit to the top of a building)	May increase the building's susceptibility to explosion damage and, therefore, increase the risk to occupants
Replacing weathered siding on the chemicals warehouse using the same materials of construction	Replacing the floor in the chemicals warehouse, eliminating the curbing between warehouse sections.	Loss of engineered control intended to prevent the intermingling of potentially incompatible liquid spills
	Increasing the size of the chemicals warehouse	Revised requirements for sprinkler protection may exceed the flow/pressure capability of the fire water supply
Repaving an existing road while maintaining existing drainage, shoulder elevation, width, etc.	Temporarily closing a major site road due to interferences from a construction project or a maintenance turnaround	Potential to impact the accessibility of emergency response vehicles to certain portions of the facility

Replacement-in-Kind	Change	Example Concern with Change*
	Changing delivery truck routing through a site	Potential for exposing offsite drivers to higher hazard areas of the facility
Making a repair to a building system in a fashion that conforms exactly to the original design specification	Replacing a building ventilation air intake stack with a shorter one	Increase the potential for the ingress of flammables or toxic contaminants into the building ventilation supply
	Replacing the impeller in the building air supply fan with a smaller impeller, or reducing the horsepower of the drive motor	May not be able to satisfy the pressure and flow requirements established to maintain the positive pressurization of the building required to prevent the infiltration of flammable or toxic materials
	Removing the weather stripping on the doors to a building within an electrically classified area	Potential for invalidating the electrical classification established for the building interior
Making a repair to a facility utility system in a fashion that conforms exactly to the original design specification	Making a temporary repair to a firewater header using lower-schedule or smaller-diameter pipe, pending receipt of specification piping	Potential for reducing the firewater header flow capacity. May need to reduce the header pressure consistent with the allowable working pressure for the temporary piping, with consequent reduction in the horizontal and vertical reach of fire monitors
	Raising the steam header pressure	Existing pressure regulators supplying branch lines or equipment may adequately maintain the lower service pressures required at points of use, but steam may be superheated (i.e., at higher-than-design temperature) after the pressure reduction
	Switching to an offsite supply of a critical site utility (e.g., steam, firewater, inert gas from a pipeline) rather than using a dedicated source under facility control	Loss of control over the quality and supply of the service

Replacement-in-Kind	Change	Example Concern with Change*
	Adding a branch line to an existing utility system (e.g., firewater, nitrogen for inerting, process water) without making corresponding increases in the supply capacity	Potential for reducing supply pressure below that required during periods of high consumption in the new branch header
Reassigning emergency response roles among equally capable personnel	Disbanding facility emergency response capabilities in lieu of support from municipal emergency response agencies (e.g., doing away with the site fire brigade and depending upon the public fire department instead)	Relying on municipal agencies to maintain the required capabilities and response times. Loss of control over the availability of specialized capabilities. Competition for available resources in the event of a natural disaster affecting multiple sites in the area
Organizational and Staffing Issues		
Replacing an employee with a comparably qualified employee (or providing suitable overlap between the incoming and outgoing employees to allow the new employee to gain the needed qualifications)	Deciding not to replace a retiring corporate loss prevention expert who was previously assigned to review all relief system designs	Loss of specialized expertise with the possibility that comparable expertise will not be available (at least in the near term). Possibility that the responsibility for reviewing relief system designs will be addressed only on an ad hoc basis and, therefore, that some designs may not be reviewed. Personnel having unofficial responsibility for reviewing relief system designs may not be given the resources, or may not take the initiative, to stay abreast of technological advances
	Realigning the corporate PSM auditing function, placing primary auditing responsibility at the site level	Site-level, part-time auditors may not have the same expertise in auditing or the same objectivity and independence as corporate personnel
	Relocating the site technical group to a remote corporate location	Technical personnel could only perform their oversight role indirectly. They could not easily interface directly with or help train operating personnel

Replacement-in-Kind	Change	Example Concern with Change*
Reassigning personnel from one shift team to another while maintaining the same basic staffing structure	Eliminating an outside operator position	Process monitoring function could be less effective. Field response to process problems could be less timely
	Changing from an 8-hour shift schedule to a 12-hour shift schedule	Need to review human factors issues related to the new shift schedule (e.g., fatigue associated with longer shifts)
	Reducing the number of operators on a shift	More tasks and greater responsibilities assigned to remaining operators. Need to clearly define the new distribution of responsibilities
	Increasing the number of operators on a shift	Need to clearly define the new distribution of responsibilities. Added administrative burden for shift supervision. More personnel exposed to process risks
	Eliminating shift supervisors and moving to a self-directed workforce concept	Need to clearly define the new distribution of responsibilities. Operators potentially unwilling to assume responsibilities previously assigned to supervision
Replacing the current maintenance contractor with another qualified contractor	Changing from centralized maintenance to decentralized maintenance	Need to address issues such as: • Responsibility for training • Standardization of procedures and practices • Potential that mechanics will spend a greater percentage of their time in closer proximity to process hazards • Need to train more mechanics on process overviews, chemical hazards, MOCs, etc.
Promoting a properly qualified operator to chief operator	Changing the qualifications for chief operator	Need to establish the basis for the assertion that the new qualifications increase (or at least do not decrease) the chief operator's ability to help control process risks

Replacement-in-Kind	Change	Example Concern with Change*
Policies		
Rigidly enforcing the requirement that face-to-face shift turnovers be conducted at the job site	Tacitly allowing turnovers to be conducted in other locations or in other ways (e.g., in change rooms or parking lots, through intermediaries)	Less effective communication of process status. Potential for problems not to be communicated to oncoming personnel
Communicating with the intent to clarify or reinforce existing policies (without modifying such policies)	Liberalizing the amount of overtime that an employee can work each month	Potential abuse of the overtime system, leading to increased worker fatigue
	Revising the facial hair policy to allow facial hair for some classes of employees whose potential need to wear respiratory protection is considered low	Potential regulatory liabilities. Potential to lessen the credibility of the facial hair restrictions for other groups of employees
	Introducing a total prohibition on smoking at the site	Potential need to be more observant of illicit smoking in areas that are less public (and potentially less safe) than former smoke shacks (e.g., in remote areas such as motor control centers)
Other Process Safety Management System Elements		
Rigidly enforcing existing safe work practice procedures	Eliminating a step in the approval of safe work permits that currently requires sign-off by the control room lead operator	Potential for reducing control room personnel awareness of work being conducted in the process area. Potential for process activities/status not being properly considered when issuing safe work permits
	Reducing the frequency of flammable gas measurements required during hot work jobs	Potential to overlook changing conditions that might make the work site less safe (e.g., increases in ambient flammable concentrations due to changes in wind direction)
	Increasing the frequency of flammable gas measurements required during hot work jobs	Failure to consider and provide the increased resources necessary to implement the change (e.g., more gas detectors and workers trained to use them)
Rotating the responsibility for facilitating incident investigations among a group of comparably qualified leaders	Revising the qualifications required for incident investigation leaders	Potential for less effective investigations due to the reduced capabilities of incident investigation leaders

Replacement-in-Kind	Change	Example Concern with Change*
Formalizing the currently implemented, but not yet documented, practices for field safety inspections of contractor activities	Eliminating the contractor safety review if the contractor has been on site within the past 6 months	Potential to overlook significant changes in the contractor safety program made since the contractor was last on site
	Modifying the contractor screening protocol to allow waiving the requirements established for past safety performance (for contractors providing unique or essential services that would otherwise fail to meet the screening requirements)	Possibility of allowing a potentially unsafe contractor on site (absent other mitigating controls)
Relocating hot work to another area within a nonrestricted ("free burn") hot work site	Reclassifying an area that currently requires a hot work permit as a "free burn" area	Need to evaluate the basis for the original classification in terms of what has changed to warrant the new classification (e.g., Are conditions in the area sufficiently static to warrant a "free burn anytime" classification?)
Changing the day of the week on which operators are scheduled to attend refresher training on operating procedures (assuming that this does not disadvantage any particular shift)	Changing the frequency of refresher training	Revised frequency may not adequately reflect the rate of change in the process (i.e., refresher training may not be frequent enough), or it may place an undue burden on the operations organization, such as requiring excessive overtime (i.e., refresher training is too frequent)

*** Even changes that are perceived to be for the better are considered changes, and they may have unintended consequences.**

APPENDIX B:

MOC SYSTEM DESIGN STRUCTURE

The level of detail and effort for any particular work activity should be based on the RBPS criteria: risk, demand for resources, and process safety culture. As risk decreases, consideration should be given to less detailed implementation options, taking into account the demand for, and availability of, resources and the culture.[8]

The following work activities may not be necessary in every situation. You should evaluate your own circumstances and determine which are appropriate for your company/facility. Also note that regulatory requirements may specify a minimum level of work activity.

MAINTAIN A DEPENDABLE MOC PRACTICE

Establish Consistent Implementation

1. Establish and implement formal procedures to manage changes, excluding RIKs.
2. Assign a job function to be the "owner" of the MOC system and to routinely monitor MOC effectiveness.
3. Define the technical scope of the MOC system so that the types of changes to be managed are unambiguous and the sources of changes are monitored.

Involve Competent Personnel

4. Define the MOC roles and responsibilities for various types of company/facility personnel.
5. Provide awareness training and refresher training on the MOC system to all employees and contractors.

6. Provide detailed training to all affected employees and contractors who are assigned specific roles within the MOC system.

Keep MOC Practices Effective

7. Keep a summary log of all MOC reviews, including the items that must be included on an MOC review form, to aid day-to-day management of the MOC process.
8. Establish and collect data on MOC performance indicators and efficiency indicators.
9. Provide input to internal audits of MOC practices based on learnings from the MOC performance indicators.

IDENTIFY POTENTIAL CHANGE SITUATIONS

Define the Scope of the MOC System

10. Determine the types of changes to be addressed in the program:

- PSM system
- Plant layout or equipment location/arrangement
- Facility and equipment
- New chemicals
- Software
- Procedures
- Process technology
- Process knowledge
- Process controls
- Chemical specifications and suppliers
- Job assignments (individual, shift, or staff)
- Organization
- Policies
- Building locations and occupancy patterns
- Other

11. Document the rationale for not addressing each type of change in the program.
12. Develop a list of areas, departments, and activities to which the MOC system applies.

Manage All Sources of Change

13. Develop a list of personnel, documents, electronic recording systems, and so forth to monitor for sources of unrecognized change.
14. For each type/category of change to be evaluated, develop specific examples of changes and RIKs for each category for use in employee awareness training to minimize the chance that the MOC system is inadvertently bypassed.

EVALUATE POSSIBLE IMPACTS

Provide Appropriate Input Information to Manage Changes

15. Consider all the types of information needed to properly evaluate changes within the scope of the MOC system. Facilities should consider developing checklists of appropriate sources of input information for reviewers to use.

Apply Appropriate Technical Rigor to the MOC Review Process

16. Ensure that the written MOC procedures include the use of an MOC review form and that the following items are addressed prior to any change:

    - Technical basis for the proposed change
    - Impact of the proposed change on safety and health
    - Authorization requirements for the proposed change

17. Use appropriate analytical techniques, including qualitative hazard evaluation methods, to review the potential safety and health impacts of a change.
18. Identify issues that must be addressed in a review commensurate with the level of complexity and significance of the proposed change, regardless of the technique used. Specify quality parameters for the review results.
19. If temporary changes are permitted, ensure that the MOC review procedure addresses the allowable length of time the change can exist and confirms (1) removal of the temporary change or (2) restoration of the change to the original condition.
20. If emergency changes are permitted, ensure that the MOC review procedure defines what constitutes an emergency change and the process for evaluating an emergency change, including an "after-the-installation" evaluation.

Ensure that MOC Reviewers Have Appropriate Expertise and Tools

21. Perform MOC reviews using qualified personnel. Facilities should provide the basis for specifying reviewer qualifications.
22. Provide a description of the disciplines that are needed for an MOC review for each type of change.
23. Involve someone in each review who is qualified in hazard analysis.
24. Ensure that reviewers have access to, and are trained in the use of, company/facility hazard/risk tolerance criteria.

DECIDE WHETHER TO ALLOW CHANGE

Authorize Changes

25. Ensure that each change is authorized by a person with designated approval responsibilities. Sometimes this function is fulfilled by an MOC reviewer; other times, the approver is independent of the MOC reviewers.
26. Develop a list of responsibilities for those authorized to approve changes.
27. Make provisions for qualified backup personnel when designated MOC authorizers are not available.

Ensure that Change Authorizers Consider Important Issues

28. Address the options reviewers have in making decisions about MOCs in the MOC procedure.
29. Provide authorizers with access to appropriate company/facility risk evaluation/tolerance criteria guidance.

COMPLETE FOLLOW-UP ACTIVITIES

Update Records

30. Update all PSI prior to startup of the change. If this is not possible, facilities should use temporary records (e.g., redlined P&IDs or procedures) and then track incomplete items regularly until they are brought up to date, reviewed, and approved. Subject facilities/activities should consider specifying a maximum length of time (e.g., 90 days) during which information should be updated after implementation of the proposed change.

Communicate Changes to Personnel

31. Communicate changes to potentially affected personnel, including contractors. Personnel involved in operating a process, and maintenance personnel and contract employees whose job tasks will be affected by a change in the process, should be informed of or trained in the change prior to startup of the process or the affected part of the process.
32. Document that the training was completed and that the employees understood the training.

Enact Risk Control Measures

33. Create a system to address MOC review action items and to document their completion. The facility should address which action items are to be completed before the change is implemented and which can be completed following the change.
34. Confirm measures to restore conditions of expiring temporary changes.
35. If emergency changes are permitted and the MOC procedure allows authorization of changes without the immediate updating of records, establish measures to ensure that the MOC procedures are completed as defined for emergency MOC situations within a designated time.

Maintain MOC Records

36. Prepare MOC review packages that contain materials and information used by reviewers and authorizers to perform the review. Review packages should be retained for a specified period (e.g., 1 to 5 years, but at least until the area's next PHA or PHA revalidation has been completed) to support other PSM work activities.

APPENDIX C: EXAMPLES OF MOC SYSTEM PROCEDURE WORK FLOW CHARTS AND MOC REVIEW DOCUMENTATION FORMS

The following figures provide examples of MOC system procedure work flow charts and MOC review documentation forms. These charts and forms should be considered advisory; they should not be used directly without going through the proper MOC system design and development activities discussed in Chapters 3 and 4.

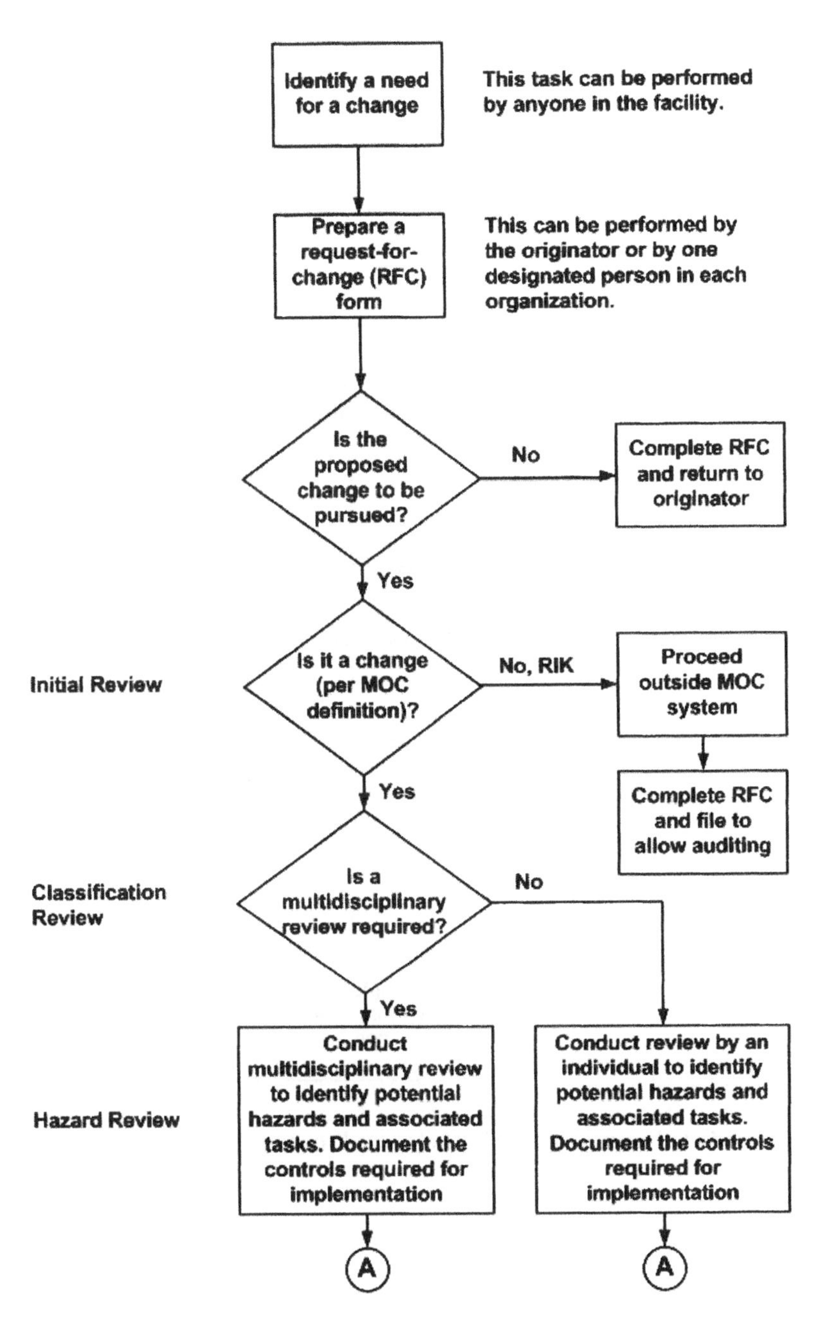

Figure C.1 Example of an MOC System Procedure Work Flow Chart

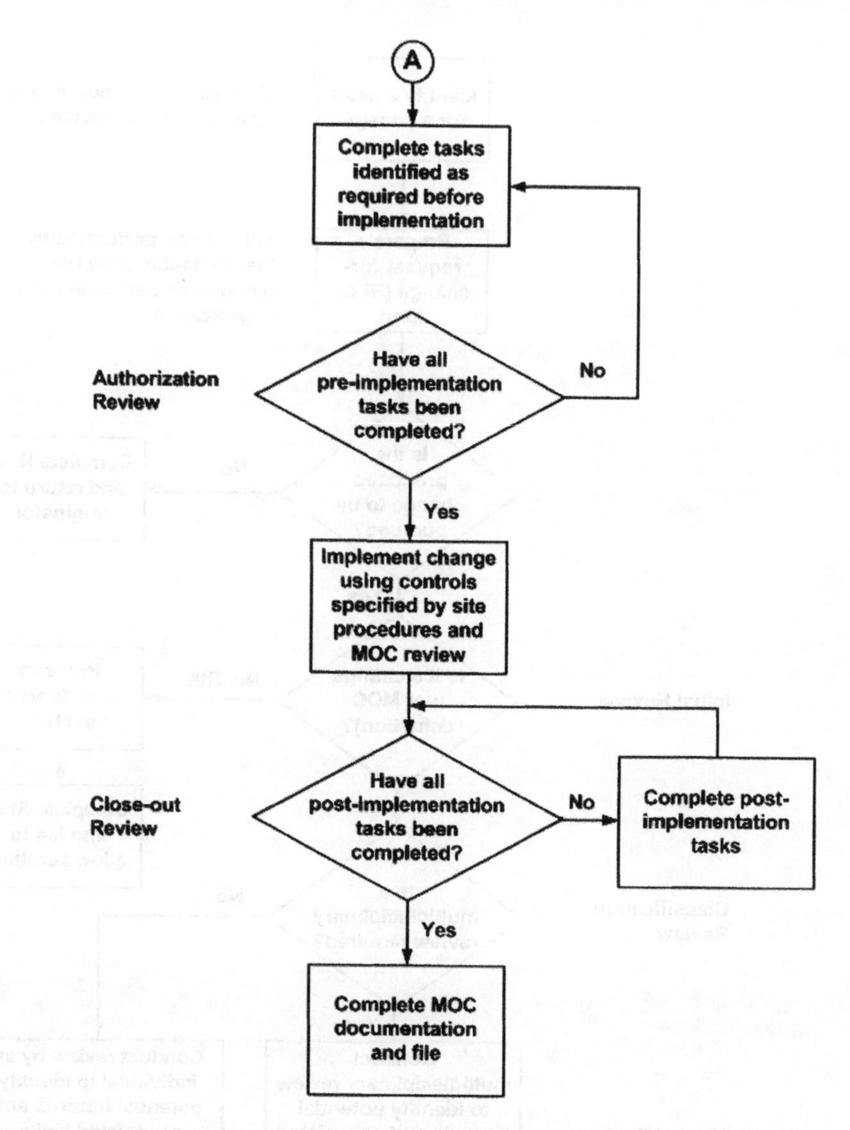

Figure C.1 Example of an MOC System Procedure Work Flow Chart (*cont'd*)

Change description and rationale:

If the change is temporary, list the pertinent dates. Dates valid:

____

Originator

This change has met the appropriate review requirements and has been approved. Safety, health, and environmental concerns have been addressed, procedures have been revised, the appropriate training and/or communication activities have occurred, and all affected process safety information is being updated.

____

RFC Authorizer

FIGURE C.2. Simple MOC Review Documentation Form

Unit or Area: RFC No.: ____________

Description and reason for change: Date: ____________

____
Originator

❑ Temporary Change Removal Date:

Environmental, health, and safety reviews are complete and all concerns have been addressed.

EH&S Review Team Leader

Operating, maintenance, and emergency procedures have been reviewed.

____
Area Procedures Coordinator

All affected personnel have been informed of the change. The appropriate training has taken place.

Area Training Coordinator

All affected process safety information is scheduled for revision.

Unit Engineer

❑ PSSR Required PSSR No.: ____________

This change has met the appropriate review requirements and has been approved.

Area Manager

FIGURE C.3. Moderate MOC Review Documentation Form

REQUEST FOR CHANGE FORM

Standard Change ☐
Emergency Change ☐
Temporary Change ☐

RFC No. ______________
Date Requested ______________
Date Required ______________

Unit ______________________ System or Equipment ______________________

Description (include technical basis for change):

____
Originator

I. Temporary Changes (skip for permanent changes). This information may be provided in a temporary procedure (attach copy)

Why is this designated a temporary change?

Additional precautions required:

Contingency plan:

Dates valid:

Person responsible for removing the change:

II. Safety, Health, and Environmental Reviews

	Req'd (Y/N)	Responsible Party	Target Date	Date Complete	Initials
Process Safety (specify method)	Y				
Occupational Safety/Industrial Hygiene					
Environmental Review (s)					

* Complete = Action items with immediate impact are resolved and plan is in place to address long range items.

III. Procedures Revised

	Req'd (Y/N)	Responsible Party	Target Date	Date Complete	Initials
Startup/Shutdown/Emergency Shutdown	Y				
Normal Operation					
Maintenance					
Emergency Response					
Other (e.g., administrative) ______					

** Complete = Revised procedures issued; any obsolete procedures discarded.

FIGURE C.4. Complex MOC Review Documentation Form (including checklists)

I. Training

	Req'd (Y/N)	Individuals or Groups to be trained	Responsible Party	Target Date	Date Complete	Initials
Operations	Y					
Maintenance						
Contractor						
Other ______________						

† Complete = All specified personnel have received and understood training. Responsibility for any change to permanent training materials (e.g., learning blocks) is assigned and scheduled.

II. Process Safety Information Revised

	Req'd (Y/N)	Responsible Party	Date Complete††	Target Date	Initials
P&ID	Y				
Process Flow Diagram					
Electrical System Documentation					
Relief System Documentation					
Spare Parts List					
MSDS					
Documented Operating Limits					
Other ______________					

‡ Complete if there are any Yes responses:

Follow-up Responsibility ______________________

PSSR Responsibility ______________________

PSSR No. ______________________

III. Authorization

This change has met the appropriate review requirements and has been approved.

______________________ ______________________

Area Operations Manager Area Engineering Manager

IV. Close-out Review

All of the indicated process safety information revisions (Section V) have been completed. This MOC action is complete.

FIGURE C.4. Complex MOC Review Documentation Form (including checklists) *(cont'd)*

Plant modification tracking no.:	❑ Off-hours review MOC initiation date ______ Required implementation date
Modification title: ______	
Permanent ❑ Temporary ❑ Expires ______	
Requestor: Date:	
Unit: ❑ FCCU ❑ HF-Alky ❑ Crude ❑ Quality Control ❑ H_2S ❑ Reformer ❑ Terminal ❑ Maintenance ❑ Other	
Estimated costs of modification: Parts, material, equipment Labor Total ______	
Description of change	Does a sketch accompany this form? ❑ Yes ❑ No
Purpose (technical basis)	
Potential impact on safety, health, and the environment (including the public)	
Reviews ______ *Area Manager* ______ *Date* ❑ Approved ❑ Denied Comments, conditions, or reasons for denial ______ Hazard review required ❑ Yes ❑ No Technique: ❑ MOC ❑ What-if ❑ HAZOP ❑ Other ______ *Initials*	
Area Engineer *Date* ❑ Approved ❑ Denied Comments, conditions, or reasons for denial Design responsibility assigned to ______ Design review required ❑ Yes ❑ No	

FIGURE C.4. Complex MOC Review Documentation Form (including checklists) *(cont'd)*

APPENDIX D:
ELECTRONIC MOC APPLICATIONS

With today's prevalent use of computer systems and their continuing evolution as an essential business tool, many companies are choosing to convert their paper-based management systems to computerized systems. Electronic MOC (eMOC) systems offer a significant opportunity to improve the performance and efficiency of an active MOC system. However, improperly executed eMOC projects can set MOC practices back years in terms of decreased performance.

D.1 BACKGROUND

Companies considering the adoption of eMOC generally start from one of four initial conditions:

No MOC system exists. The facility is starting from scratch and develops its MOC procedures and work flow while implementing eMOC.

A paper based MOC system exists. A company wants to transition from a paper based system to an eMOC system, probably due to being overwhelmed by MOC activity, increasing backlog, and unreliable results and follow-through.

An e-mail based MOC system exists. The facility has gone part of the way: it has a paper based system, but it passes the paper along electronically via e-mail attachments. In some cases, MOC training may be conducted/ managed through the same e-mail system.

A failed eMOC system needs to be revived or replaced. A previous attempt to implement an eMOC system failed, probably because personnel did not have access to or did not follow the type of advice presented in this appendix. The company wants to try again.

No matter what the starting point, the business decision to computerize an MOC system should address two issues:

1. Decide to improve MOC performance, efficiency, or both by moving from a paper-based MOC system to an eMOC system.

The major advantages of an eMOC system involve tracking, documentation and compliance, automatic routing, and automatic reminders. By automating these functions, the chances for human error can be reduced. Furthermore, developing an eMOC system can provide the opportunity to standardize the MOC process, which offers more room for improvement. Also, collecting data is much easier with an electronic system, and using measures and data to manage and improve the system is also simpler. An electronic system can also aid assessments and auditing.

The primary challenges of implementing an eMOC system concern issues relating to technology (hardware/software considerations), training, and communications. Going from a paper based system to an electronic system is a major shift, and it can disrupt normal communication patterns in unexpected ways (especially if people depend upon meetings and other face-to-face communication as they did under the old system). Table D.1 lists some benefits that are typically anticipated when deciding to implement an eMOC system.

2. Determine whether the new eMOC system should be developed in-house or through the purchase of a commercial product.

In-house development of eMOC systems may be attractive to companies that are experienced in developing web-based or server-based software applications. Creating the application internally has the following advantages:

- Personnel are very familiar with company/facility MOC needs
- Can dovetail eMOC system more easily with existing applications
- Internal personnel are available to consult with plant personnel if modifications/upgrades are needed

TABLE D.1. Anticipated Benefits of Choosing the eMOC Approach

- More dependable work flow with programmed feedback loops and recycle of reviews/authorizations
- Improved communications flow, tracking and automatic routing, less bottlenecking, automatic reminders
- Reliable records, archiving of approvals, documenting accountability
- Electronic time stamps to prevent postdating of approvals or sign-offs or writing MOCs after the fact
- Less need for meetings of MOC reviewers – compensate using virtual/net meetings
- Ease of use, faster reviews/approvals, potentially more cost effective
- Easier auditing/metrics generation

However, for many companies, the internal resources needed to build an eMOC application either do not exist or are insufficient; therefore, the decision is made to purchase a commercial product. Even if resources are available to design and build a custom eMOC application, careful consideration should be given to future system maintenance needs prior to embarking on eMOC application development.

Whether developed in-house or purchased, however, the process for implementing the application is the same. One cannot simply take a work process based on a paper system and apply it to an electronic system. If this is attempted, the company will end up with a manual system that resides on the computer without any of the anticipated business improvements. The resulting system will simply serve as a way of easily accessing the required forms. An effective eMOC system has no forms to download and stores the information in a database.

D.2 DESIGN CONSIDERATIONS

Several important considerations exist for those interested in either internally developing their own eMOC system or purchasing one from a commercial vendor:

- ***Standardization versus the flexibility to account for differences in manufacturing areas.*** Software applications have the advantage of ensuring that the appropriate MOC workflow is used each time an MOC is reviewed/approved. If a plant needs different approaches to be used in plant areas, the software can be programmed to recognize and implement these different MOC review protocols. However, the more variety that exists in MOC protocols, the more tedious it is to control the implementation of eMOC systems.
- ***Layers of protection in MOC approval workflow versus ease of use.*** Various confirmation steps can be included to add layers of protection in the MOC review/approval process, but these also complicate the process and

make it more difficult to use. This is analogous to implementing electronic process control systems. In a digital electronic control system, alarms are easy to implement (compared to an old analog control system); so putting alarms on everything can be tempting. However, if this is done, the operator can suffer from alarm overload and miss the critical ones.

- ***Streamlining work flow steps.*** Work flow steps are needed for critical tasks. However, a separate work flow step for each task may make the process too complicated. One solution is to combine several tasks into one work flow step, assuming that one person (such as the originator) can be assigned responsibility for getting those tasks done (although this person doesn't necessarily have to perform all the tasks). Another streamlining option is to simply notify appropriate personnel about a change rather than including this as an actual work flow step. Generally, a simpler work flow process in which some tasks are performed "off-map" may be better received.
- ***The role of the originator.*** Giving the originator responsibility for more tasks keeps him or her involved in the MOC process, but also requires more of his or her time. Also, the originator may not be an expert on the MOC process. Alternatively, responsibility for some of the steps can be given to a central coordinator or coordinators who are experts on certain aspects of the MOC process. However, these people will probably be less knowledgeable about the details of the request.
- ***How detailed to make the checklists.*** A more detailed checklist has the advantage of reducing the chance that an important consideration will be overlooked (in other words, the process is less dependent upon the expertise of the user). However, a more detailed checklist has the disadvantage of requiring more time to complete.
- ***Making the system flexible enough to handle the many different types of changes.*** This can possibly be managed through a combination of mandatory and optional work flow steps. For very simple and lower-risk changes, some work flow steps may not be needed and can be skipped, resulting in a simplified work flow.
- ***Simplicity versus functionality.*** The MOC system should be as user friendly as possible. The ideal is to have an electronic fill-in-the-blank form and to give those performing the work flow tasks only the information they need (i.e., not overwhelming them with extraneous information). Also, the system should be intuitive, thereby not requiring a high level of computer expertise to use it. If the eMOC system is more difficult to use than the paper based system, people will resist using it, even if they see the advantages. Special consideration should be given to the casual user who may access the system only occasionally or who may have a limited role within the system, such as reviewer or approver.
- ***Ability to attach supporting documents that can be quickly displayed.*** Some plants have change types that require extensive backup analysis documentation and data. In these situations, the eMOC system should be

capable of associating these documents and data with the MOC review form as it goes through the eMOC workflow. At various stages, more records may be appended to the MOC record, and these documents must all be accounted for in archiving the MOC results.

- ***PC-based vs. LAN-based vs. internet-based.*** Some plants may have a preference for the computer platform/environment upon which to implement the eMOC system. The selection of the platform should anticipate the rate of use, speed, and resource requirements of the anticipated computer environment.

If a commercial product is chosen, additional factors should be considered. Many commercial products incorporate more than just an MOC application, so every effort should be taken to understand not only what is included in the product, but also how each module of the MOC application interfaces with the others.

Other critical considerations are: how compatible the commercial MOC application is with existing systems and how easy it is to adapt to changes in other systems. For example, the MOC application will need to interface with the existing work order system and engineering applications. In order to ensure the effectiveness of the eMOC application, functional and technical specifications will need to be developed.

How data will be filed and retrieved is an important consideration. Data are best stored in a single location, such as a database, so that they can be automatically filed and easily retrieved. This limits where data can be stored and ensures that data are not misfiled or lost. This also provides a robust audit trail. Therefore, system designers should determine which data are mandatory and which are optional so that necessary data are available when queried.

Another consideration is the use of electronic signatures for approvals. Appropriate levels of security can be built in to ensure that the appropriate authorization has been granted without obtaining hard-copy signatures. If a facility chooses not to have this feature, then some form of hard copy will be required and this will need to be included in the specification.

How MOC data will be routed and tracked also needs to be addressed. The most convenient method is by automatic e-mail alerts to those who need to work on, review, amend, or approve a specific change. Also consider including a status tracker in the system so that, at any point in time, the originator can find out where a specific MOC is within the process.

D.3 BUILDING AN eMOC SYSTEM

If a company decides to build its own eMOC system, the following general eMOC implementation steps should be considered:

Step 1: Design the work flow
Step 2: Develop a functional work scope
Step 3: Build the prototype
Step 4: Conduct multiple tests
Step 5: Make final revisions
Step 6: Conduct training
Step 7: Roll out the system and provide support

Step 1: Design the work flow

The work flow needs to accomplish two things: (1) it should identify each step in the process and (2) it should show how each step is linked to the next through a go/no-go path.

Start by identifying each basic step in your work flow; determine why it is necessary and ensure that it is linked appropriately. Figures D.1 and D.2 illustrate generic MOC work flows (simple and detailed, respectively).

Step 2: Develop a functional work scope

The key to developing a good functional work scope is to first develop clear definitions if they do not already exist. For example: Will the application be used for non-engineered changes, such as administrative, organizational, and personnel changes? Also, will the eMOC system be applied to any change within these categories, or just for critical positions or personnel? Clearly defining the types of work activities that would not be considered changes is also important. This ensures that personnel will understand the circumstances under which to initiate an MOC.

Developing a functional work scope provides the application designer with a detailed blueprint of the needs and expectations for the application. Specific information is required for each segment of the work flow. For example, for the segment "Open MOC", the issues listed in Table D.2 would apply.

Whether a company is building the application in-house or using a commercial application, developing a functional specification is a critical step. Most commercial applications are adequately robust and have been designed to allow customization to fit a company's business language and work flows. Commercial applications have generic, universal work flows and templates that can be used. However, some established work flows may have to be altered to fit the application.

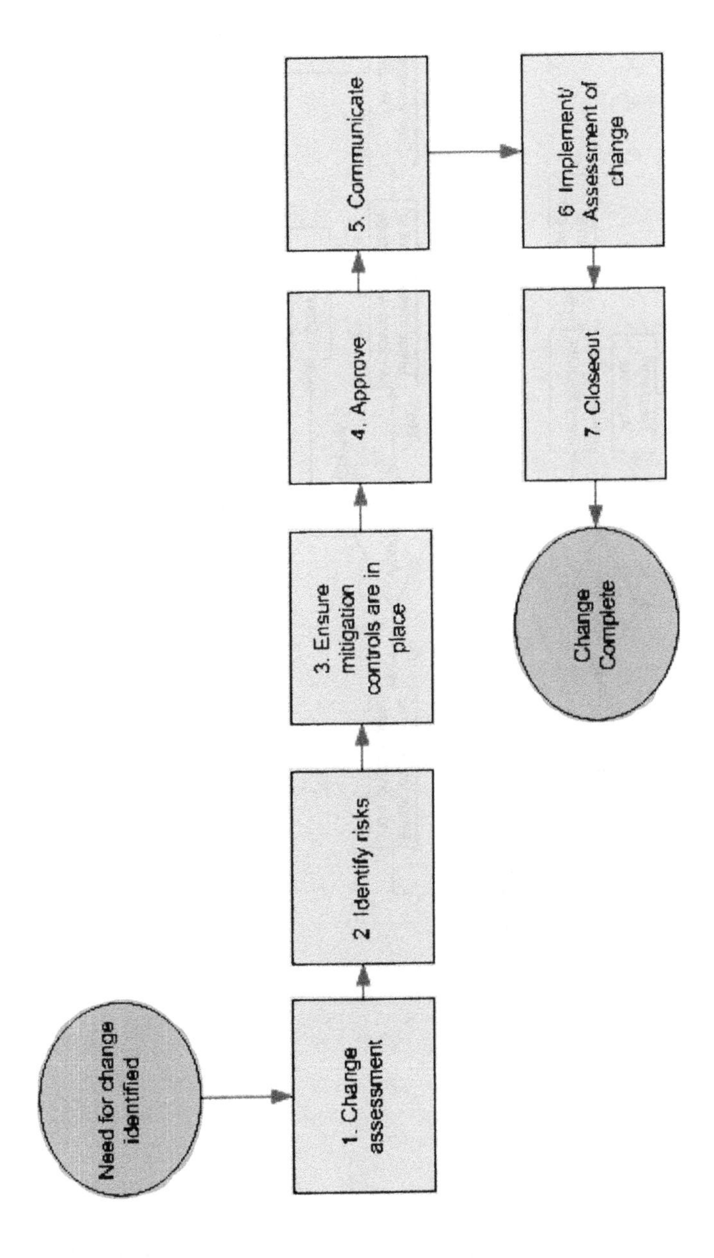

FIGURE D.1 Simple Generic MOC Work Flow

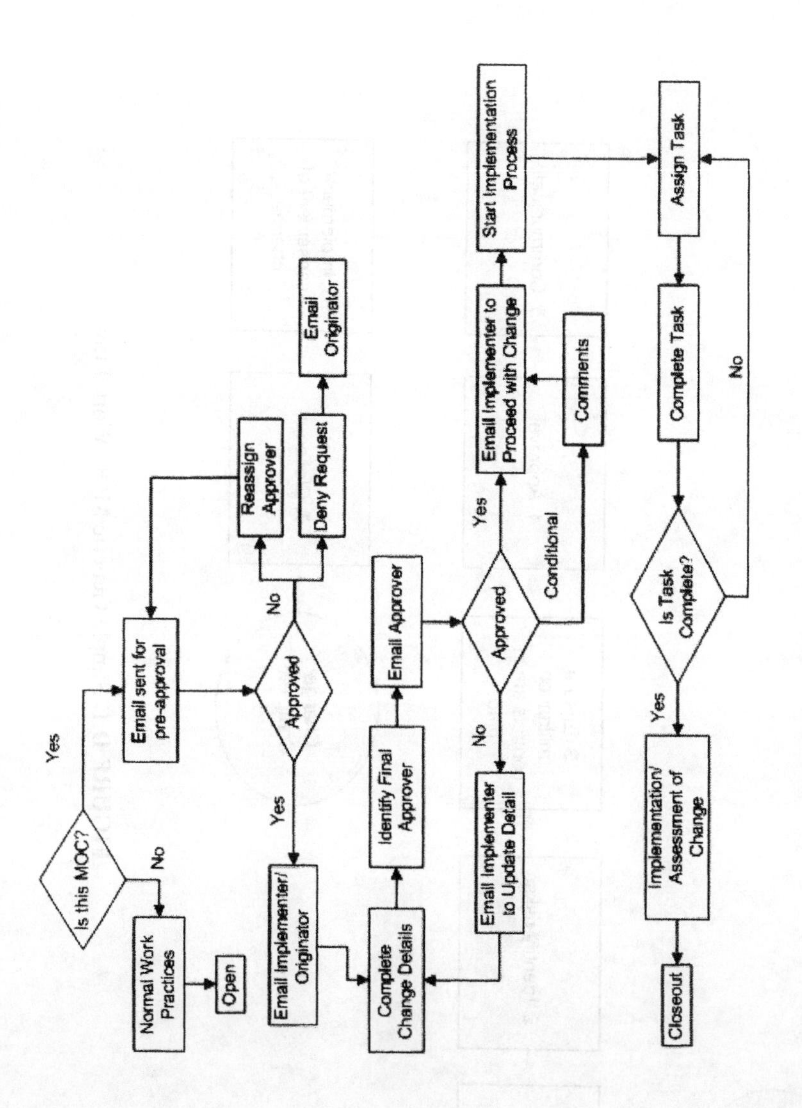

FIGURE D.2 Detailed Generic MOC Work Flow

TABLE D.2. Open MOC Template

Field	Description
Originator name	Text box
Department	Pick list (data table)
MOC number	Application generated
Date	Automatic
Change category	Pick list (data table)
Change type	Pick list (data table)
Brief description	Text field
Pre-approver name	Data table
E-mail approver	Automatic by system

The process should be designed to ensure that appropriate communication takes place among key participants, which includes those doing field validations and those who should be notified of a change. This will provide the developer with enough information to build templates for review. Critically review these templates and make any required changes at this point. Having sufficient rigor in design reviews at this stage will reduce recycling in future steps.

Step 3: Build the prototype

At this stage, the application developer will ensure that all of the fields on each template function correctly and that the application is user friendly. Engage personnel who will use, review, and critique the application. At this stage, wording choices and data field order are often changed to eliminate confusion or multiple interpretations.

Step 4: Conduct multiple tests

This step is accomplished by having a number of users test the application, ensuring that all types of users are represented. Generally, users fall into one of three categories:

- *Power users.* Those who will use the application frequently and are in the best position to judge its effectiveness and functionality.
- *Casual users.* Those who will use the application occasionally and will need reminders on how to use it.
- *Approvers.* Those who will use only one aspect of the application.

Have each tester follow the process by entering data and putting the application through its paces; ensure that users test each function. Have each user provide both positive and negative feedback.

Step 5: Make final revisions

Use the test feedback to make final adjustments to the application before it goes into production. If a commercial product is being tested, this stage is where any compatibility issues with existing applications will need to be discovered and addressed before the system is rolled out. This step verifies that the application will work as specified.

Step 6: Conduct training

Don't underestimate the amount of training that will be required. First, identify the extent of the target learner's knowledge of the eMOC system, its purpose, and how it is used. Then build the appropriate training. Two levels of eMOC training will be needed: (1) awareness training for all personnel on the system and its application and (2) specific training on its use.

Training should be conducted on computers, allowing each participant to actually use the application. This may involve a number of activities. For example, the approvers of changes are generally managers who won't want to sit through training on how to use the entire application. They will need a more targeted training program that meets their specific needs. Include training for the help desk personnel or other support staff. All training should be conducted before the application is rolled out.

Step 7: Roll out the system and provide support

Use the company's current process for introducing new software and communicate that the eMOC system is now available. Emphasis should be placed on system support, as there will be lots of questions from users once the application is rolled out.

D.4 PURCHASING AN eMOC SYSTEM

If a company decides to purchase an eMOC system, it should start by developing a technical specification outlining the needs and constraints for the system. Depending upon internal policies, a company might either (1) purchase the software and then perform its own installation or (2) purchase the software from an application service provider (ASP). In either event, the application should be compatible with the company's specific hardware and other systems. Table D.3 provides some questions that should be asked when developing a technical specification to provide to a potential commercial product supplier.

Providing the following information is essential when requesting information from potential MOC software suppliers:

- Detailed description of the application package(s), focusing on core functions. Include discussion of any third-party application tools that are being recommended to address both functional and technical requirements.
- High-level project plan based on estimated resources assigned to the implementation project, accompanied by a description of the defined roles and the skills required for each role (including client involvement)
- Description of the network, server, desktop, and architecture, if the vendor concludes that an alternate technical architecture is required to support the package(s)
- Detailed cost estimates for the implementation project
- Cost estimates of hardware upgrades, replacement, or additional components necessary to enable implementation of the application
- Cost estimates and timelines for technical and user training
- Cost estimates for optional enhancements recommended by recipient vendors
- Estimates of how and within what time frame each phase of the project would be implemented, broken down by subcomponent

Many factors affect the success of selecting eMOC software from a supplier:

- Cross-functional design team, including representatives from all key areas and a project manager to lead the team
- Management support, including line management and a steering team at the appropriate senior management level
- Objectives and performance measures agreed upon before the design process starts
- Data from the pilot tests used to make modifications before the tool is implemented across the board in manufacturing processes
- Core group of knowledgeable trainers
- Startup support
- Implementation coordinator named for each major area (site/division), whose duties include involving all appropriate area personnel and scheduling the training
- Investigation of technical considerations up front for both software and hardware to identify the most appropriate platforms and potential pitfalls
- Assignment of sufficient information technology resources
- User-friendly design requiring minimum workload

Table D.3 provides a checklist of issues to consider when developing a technical specification for use in eMOC software selection.

TABLE D.3. Questions to Address When Developing a Technical Specification

No.	Question	Musts	Wants
1.	Does your organization provide onsite support during implementation and deployment of the MOC application? Do you have a standard plan? Please provide some detail.	☐	☐
2.	Do you provide training for system administrators?	☐	☐
3.	What user training do you offer? Please describe the types of training available.	☐	☐
4.	Do you provide data integration services to tie your application into existing in-house information systems? What resource base do you have to provide this service?	☐	☐
5.	How do you integrate and/or customize existing content onto your product platform?	☐	☐
6.	Are you having "help" options built in?	☐	☐
7.	Do you provide technical support for your application? Please explain how this is accomplished.	☐	☐
8.	Does your application retain any history? How is this accomplished?	☐	☐
9.	How do you license your product (ASP versus local server installation)?	☐	☐
10.	Does your product provide templates for content development? Please describe to what extent. Is a third-party template available for the tool?	☐	☐
11.	How are risk assessments facilitated?	☐	☐
12.	During the past 2 years, what has the frequency of maintenance service been for upgrades and revisions?	☐	☐
13.	Please describe any broadcast-messaging features. Does the product generate open action reminders?	☐	☐
14.	Does your product have the capability to produce a range of standard and ad hoc reports? What are they?	☐	☐
15.	To what extent can additional standard reports be created without programming skills?	☐	☐
16.	Are NT, UNIX, and server base solutions required? Can the application be installed in an Oracle (or other) database environment?	☐	☐
17.	Does the application use an open platform whereby all IT infrastructures are supported and integrated? Are there any limitations?	☐	☐

TABLE D.3. Questions to Address When Developing a Technical Specification (*cont'd*)

No.	Question	Musts	Wants
18.	To what extent can you provide connectivity to third-party systems/products?	☐	☐
19.	What are the connectivity options?	☐	☐
20.	Is the application supported in Internet Explorer?	☐	☐
21.	What intranet, Internet, or portal environments are supported? Are there any restrictions?	☐	☐
22.	Which databases do you support and which database is your application built on?	☐	☐
23.	Will your product work with our company's existing PCs?	☐	☐
24.	Is any software installed that would limit data access?	☐	☐
25.	Can multiple users make concurrent updates? Are there any restrictions?	☐	☐
26.	Does your product provide user-based and file-based access control (for security purposes)? Please provide some detail.	☐	☐
27.	Does your product have any features to facilitate searches? Is it possible to enable or include a feature for key word searches for content?	☐	☐
28.	Does your product have any other key features related to management of change that are not addressed by these questions? If so, please specify.	☐	☐

An important consideration during rollout of the eMOC system is ensuring proper communication among key participants. As previously noted, going from a paper based system to an electronic system is a major shift, and it can disrupt normal communication patterns in unexpected ways (especially if people depend upon meetings and other face-to-face communication as they did under the old system). Reviewing documents (possibly including drawings) on a computer instead of on paper also requires a significant adjustment. The plan to ensure proper communication could include continuing with face-to-face meetings, unless they are definitively shown to be unnecessary. In other words, don't assume that the electronic system will take care of all needed communication.

A plan should be in place for ongoing management and continuous improvement of the eMOC system after it has been implemented. An individual (or a position) should be assigned as owner of the system, and the management plan should include appropriate training and performance measures.

Table D.4 lists a number of other factors that should be addressed to ensure success.

TABLE D.4. Other Important Issues Related to eMOC Systems

- Communication/integration with all electronic systems (e.g., e-mail and other administrative systems). The eMOC application should not stand alone
- Level of integration when considering off-the-shelf software
- Workforce training issues (e.g., computer accessibility and literacy)
- System dependability (e.g., backup servers, data links, emergency procedures)
- Types of eMOC systems (e.g., off the shelf, electronic data management system (EDMS) based, custom developed)
- IT infrastructure environment (i.e., PC based, client server based, or Web based)
- Design parameters (considered up front), such as rate, volume, types of DCSs, EDMS capability, and speed
- Verification/improvement of MOC work flow throughout the process (will fail if people see it as a bottleneck rather than a tool)
- Access control, passwords, communications, and sign-offs (electronic signatures)
- Multiple databases and hard-copy backup of records when archiving
- Tracking of initial and long-term time investments (prove benefits verses long-term monitoring of efficiency
- Pilot tests (e.g., table top, full exercise, full feature)
- Adequate training prior to rollout
- IT department involvement from the beginning (even if IT personnel are not developing/customizing it)
- Involving a range of users (not just senior management) from the beginning (consider establishing a user steering committee)
- Geographical and cultural considerations (e.g., multiple domestic sites, multiple international sites, multiple languages)
- Level of computer access granted to contract employees

APPENDIX E:
EXAMPLE MOC SYSTEM AUDIT CHECKLIST

Companies use audits as one way of determining the health of a management system. Some very active management systems – such as MOC systems – involve frequent work activities and generate regular work products. These types of management systems are good candidates for using performance indicators to monitor the health of the system on a near real-time basis. However, companies have other reasons for conducting management reviews and periodic audits of its management systems. Sometimes these activities are part of the company's continuous improvement processes, but nearly all companies conduct MOC audits to assess the system's conformance with regulatory requirements and/or company standards.

This appendix provides some suggestions for conducting audits of MOC systems, either independently or as part of a broader PSM or environmental, safety, and health audit. The amount of effort spent on conducting an MOC audit will be based on (1) the level of rigor applied when selecting and implementing process safety activities for this element and (2) the MOC system's activity rate (i.e., the number of changes evaluated each month or per year). This appendix describes areas of inquiry to pursue when determining whether the process safety activities are being implemented as intended (i.e., as described in the MOC system).

Audits of MOC systems should be performed periodically to help ensure that procedures described in system documents are actually being implemented in the field. The exact items to be addressed during the audit depend upon a variety of factors, including (1) the specific MOC system design, (2) the availability of MOC records, (3) the frequency of MOC reviews at the site, and (4) the period of time since the last audit.

The possible areas of inquiry are discussed in this appendix according to the three standard auditing techniques:

- Document review
- Interviews
- Field observations

The discussion takes the form of questions to consider asking when developing an MOC system audit protocol. Some suggestions are given in terms of MOC work activity/product sampling that can be performed to ensure adequate thoroughness. The audit protocol should also address other factors, such as the availability of audit personnel, the culture of the company/site, and regulatory concerns – topics that are not addressed in detail in this book, but that are discussed more thoroughly in other CCPS and industry publications.

Document Review

1. Is there a written program describing the MOC system? Does it specifically address roles and responsibilities, scope, activities, authority, and necessary documentation?
2. Does the MOC system address a reasonable range of types of changes for the facility/activity for which the MOC system is used?
3. Are the following issues specifically addressed in the MOC system?

    - Technical basis for the proposed change
    - Safety and health considerations associated with the proposed change
    - Authorization requirements for the specific class of change
    - Informing or training potentially affected personnel
    - Updating relevant process documentation and procedures

4. If temporary changes are allowed, does the MOC system address the following issues?

    - Maximum time period during which the change can exist without further review
    - Special monitoring required for the proposed change
    - Explicit field verification that the change and any associated special conditions are discontinued at the end of the time period allowed for the change
    - Adherence to time extension rules for the change

5. If emergency changes are allowed, do the requirements of the emergency change procedure meet the minimum MOC system requirements?

    - Are specific means addressed for ensuring that affected personnel are trained prior to their involvement with the change?
    - Is there an interim approval process with subsequent completion of the formal MOC review process?
    - Is there an explicit mechanism for ensuring that affected documentation is updated (if needed) in a timely fashion?

6. Is MOC effectiveness considered in the performance reviews of people who participate in the MOC system?

Scrutinize a representative sample of the MOC records for each site area in which the audit is performed. The following issues should be addressed:

7. Are the documents complete? Is there a pattern of information missing from the records?
8. Do the change requests contain all of the proper authorizations?
9. Were all of the required reviews/analyses performed?
10. Are all appropriate supporting documents appended to the MOC documents?
11. As indicated by the MOC documents, were the analyses of safety and health considerations of adequate quality, thoroughness, and depth, considering the risk significance of the change?
12. Were all affected procedures (e.g., operating, maintenance, emergency) updated by the specified time (prior to or after the change, as authorized)?
13. Were all affected drawings (e.g., P&IDs, area classifications, equipment/facility arrangement maps) updated by the specified time (prior to or after the change, as authorized)?
14. Are there any anomalies apparent with the times/dates associated with the reviews, authorizations, and start-ups?
15. Was the emergency change review procedure used frequently? Is there a trend?
16. Was the emergency change review procedure used appropriately?
17. Have there been any documented failures of the MOC system?
18. Have any change situations not been reviewed by the MOC system, as evidenced by the following types of surveys/inspections?

    - Alarm, interlock, or safety system bypass logs
    - DCS change logs

- Engineering work requests
- Revision dates on P&IDs and procedures
- Shift logbooks
- Incident investigation results
- Procedure reviews/certifications
- PHA team reviews
- Periodic walkarounds/safety inspections
- Interviews with operating and maintenance personnel

19. Scrutinize a representative sample of work orders/maintenance requests, capital change requests, P&IDs, and procedures on file for each site area in which the audit is being performed, and address the following issues:

    - Does the proper MOC documentation exist?
    - Can changes to the P&IDs be traced back through an MOC request?
    - Can changes to the procedures be traced back through an MOC request?

20. Review personnel records, organizational charts, and other appropriate documentation to determine whether any changes in the number of personnel, shift/crew size, personnel physical location, or reporting/communication relationships have occurred (pay particular attention to personnel changes that have occurred over the past 1 to 2 years).
21. Did personnel newly assigned to the facility receive MOC training, and was this training documented? Did the training include general information on the site's PSM program and policies, specific process hazards, and layers of protection, and more specific information on their roles and responsibilities within the MOC system?
22. Does the site have formal criteria or guidance that addresses the maximum rates of change for personnel in operator and mechanic roles over a specific period of time? Consider the impacts of transfers, retirements, work force reductions, leaves of absence, and reorganizations.
23. Does the site training program include, at a minimum, the following site and/or area key PSM/MOC-related roles?

    - Line management (from frontline supervisors up to and including site managers)

- Technical (e.g., technology guardians, area process engineers, R&D chemists)
- Maintenance/reliability engineers
- PSM coordinators
- Planners/schedulers
- Contractor coordinators
- DCS/process control resources
- Equipment inspectors/nondestructive testing personnel
- Resident contractors (e.g., supervisors, engineering designers)
- Operators
- Mechanics

24. How does the site determine the competency of newly assigned personnel (e.g., field demonstrations, written or verbal testing, panel reviews)?
25. Have there been any recent significant changes in the site organizational or functional structures and, if so, how were potential MOC issues considered and addressed during these changes? How is this documented?

Interviews

Perform interviews with site personnel responsible for using the MOC system (e.g., operations, maintenance, engineering, safety), and determine the following:

26. Are they aware of the MOC procedures?
27. Do they know what a change is? An RIK?
28. What is their role within the MOC system?
29. Have they received the appropriate MOC system training?
30. Can they explain the basics of the MOC procedures?
31. Do they know who can approve changes?
32. Do they know who can originate a change request and how to originate one?
33. How are they notified of a change?
34. Do they know how to have changes approved during an off-shift?
35. Do they believe that the MOC system is being reliably implemented?
36. Do they have personal knowledge of any failures of the MOC system (i.e., changes that have been implemented without appropriate review)?

37. Have they received any process-specific training as a result of a specific change?
38. Was the training conducted before they had to interact with the process change while on the job?
39. Was MOC effectiveness considered in their most recent job performance review?
40. What problems have they personally noticed with the MOC system?
41. Can they describe several examples of changes they know have been made recently?
42. What would they do if they noticed a problem with the MOC system?
43. Did personnel newly assigned to the facility (within the last 1 to 2 years), and who have PSM support roles, receive MOC training? Did the training include general information on the site's PSM program and policies, specific process hazards, and layers of protection, and more specific information on their roles and responsibilities within the MOC system?

Field Observations

Select a representative number of changes recently made across all of the MOC category types and in a variety of operating areas, and confirm the following:

44. Is the equipment arrangement/installation in the field consistent with the equipment specification and the approved change?
45. Do the updated P&IDs actually reflect the field installation?
46. Have isometrics and other diagrams used for inspection purposes also been updated?
47. Do equipment specifications in the official files match the equipment items in the field (e.g., data sheets match the nameplates)?
48. For "new and shiny" installations observed in the field, can such installations be traced back to verify that the MOC reviews were completed (assuming the work was not RIK)?
49. Do the emergency changes selected for review meet the facility definition of an emergency?
50. Were the temporary changes selected for review returned to the original condition prior to the expiration date for the temporary change?

APPENDIX F:

EXAMPLE MOC PERFORMANCE AND EFFICIENCY METRICS

This appendix discusses possible MOC metrics in light of PSM element performance and efficiency issues presented in published guidelines and experienced through industry practices.

MOC metrics that explicitly identify key indicators can be used to assess system performance and efficiency on a near real-time basis and with a reasonable effort. Below are several performance and efficiency indicators that may be relevant to many MOC systems. Monitoring these key indicators can help detect deviations within the MOC system – before they cause accidents. The sensitive indicators for a specific MOC system will depend upon a variety of factors, including the MOC system design and the availability of MOC records and data. Some indicators can be used individually to help evaluate system performance and efficiency, while others should be used jointly.

MOC Performance Indicators

- Number of incidents having MOC failure as a contributing factor or root cause
- Unexplained deviation from previous monthly averages in the number of MOCs (percent over a month)
- Unexplained deviation from previous monthly averages in the percentage of work requests classified as changes by the MOC system monitor (percent over a month)
- Percentage of work orders/requests that were misclassified as RIKs rather than as changes, or were not classified

- Percentage of changes within the MOC system that were reviewed incorrectly
- Percentage of MOCs that were reviewed but were not properly documented
- Percentage of MOCs for which the PSI was not updated
- Percentage of MOCs for which training of affected personnel was not conducted
- Ratio of identified undocumented changes to the number of changes processed through the MOC system
- Percentage of recent changes involving alternate MOC reviewers
- Percentage of changes that were properly evaluated but did not have all of the required authorization signatures on the change control document
- Percentage of changes that were processed on an emergency basis
- Variation in the percentage of changes that were processed on an emergency basis
- Percentage of temporary changes for which the temporary conditions were not corrected/restored to their original state by the deadline
- Percentage of personnel involved in the MOC system who believe it is effective
- Difference between the percentage of senior managers and the percentage of routine users who believe the MOC system is effective

MOC Efficiency Indicators

- Number of MOC reviews each month
- Number of MOC reviews in each facility/activity area each month/per year
- Average amount of calendar time between MOC origination and authorization
- Average amount of calendar time between MOC authorization and closeout of all action items
- Average backlog of MOCs/active MOCs
- Average number of man-hours expended per MOC from the time it is originated until it is approved for implementation

APPENDIX G: COMMON MOC PROBLEMS AND PROPOSED SOLUTIONS

This appendix discusses some problems commonly seen in industry that are associated with dysfunctional MOC systems. A possible solution and comments (if applicable) follow each problem description. Note: Site-specific circumstances may dictate a solution other than the one proposed here.

Problem:	One-size-fits-all MOC forms and procedures seem cumbersome or inefficient for some less common or more specialized types of changes (e.g., procedure revisions, DCS software changes). Forms geared toward equipment and/or process changes just don't seem to work well for these types of changes.
Possible Solution:	Provide a limited number of specialized forms for the more common types of changes, geared toward meeting the unique requirements of these changes.
Comments:	One CCPS member company uses a simplified approval routing for those changes that do not need to go through engineering design.

Problem:	The MOC system is not efficient for small facility changes (too complicated).
Possible Solution:	Provide a simple, low risk MOC system with a simplified work flow. It must have good guidelines, and a qualified approver (gatekeeper) must be part of the system in order to determine which MOCs qualify as simple low risk.
Comments:	None.
Problem:	There is confusion and/or indecision about the appropriate level of hazard review to perform for a given change, including the selection of a technique. There is a tendency to either perform a trivial review for a complex change or overanalyze a trivial change (which wastes time).
Possible Solution:	Provide guidance for a risk-based determination of the level of rigor to apply and the appropriate technique to use. While this presupposes, to a degree, the results of the hazard review, a conservative matrix approach has proven workable. An example might be a matrix that looks at the type of change (consequence surrogate) and the existing safety systems impacted (frequency surrogate) to determine the potential risk significance of the change.
Comments:	None.
Problem:	It is difficult to track action items that are required to be completed prior to implementing the change.
Possible Solution:	Include an action item list in the design of the MOC approval form.
Comments:	Such items should be tracked in a fashion similar to the way that PHA or incident investigation recommendations are tracked (possibly using the same system).

Problem:	It is difficult to get all of the required authorizations prior to implementation of the change.
Possible Solution:	Above all, this indicates that there is a potential process safety culture issue that must be addressed. Site management should not tolerate the startup of a change prior to obtaining the necessary authorizations. From a more tactical standpoint, include date fields next to the authorization signature blocks, and insist that they be completed. A reviewer or approver who might be tempted to sign the form after startup may not be as willing to falsify the date.
Comments:	None.
Problem:	Reviews of the potential safety and health effects of the proposed changes are not very thorough, and some significant problems have slipped through the system.
Possible Solution:	Administer these reviews as you would a PHA. Depending upon the potential significance of the change, a team-based review may be required to get the right mix of expertise. The rules used to ensure the effectiveness of PHAs should apply here as well.
Comments:	Simple, standardized changes may be evaluated by a single person using a checklist approach.
Problem:	There is inconsistency in the quality of reviews performed to determine the potential safety and health effects of the proposed change.
Possible Solution:	Many organizations use their skilled PHA facilitators to (a) lead the hazard review, (b) serve as a resource to the hazard review leader, or (c) peer review/audit the results of the hazard review.
Comments:	None.

Problem:	A lot of time is spent training personnel on trivial changes. How can we be more efficient?
Possible Solution:	The PSM standard, for example, states that certain personnel will be "informed of, and trained in, the change…" Many organizations have implemented a system whereby they distinguish between those changes about which they will inform personnel and those changes for which they will provide detailed training. A reasonable test might is to ask whether a new, fully trained operator has the knowledge needed to adapt to the change. For example, would normally be acceptable to inform operators about a change from a gate valve to a ¼-turn ball valve, but training would be required if the new valve is part of a new process operating step.
Comments:	Confirmation of understanding is commonly documented for <u>training</u> but not for <u>informing</u>. Some organizations use e-mail notifications for informing staff of changes of this nature.
Problem:	The plant has a problem with ensuring the training on MOCs for personnel who (1) are absent (due to disability, vacation, etc.) or (2) substitute for someone in a job they previously worked, but have not been involved with for a long time.
Possible Solution:	Many organizations indicate on training record forms that such personnel will be trained on the change when they return and before they first operate the modified process/equipment. Another approach is to maintain a required reading/training log in the control room and require operations personnel to check for any new MOCs at the start of each shift.
Comments:	This requires discipline and follow-up to ensure that it actually happens.

Problem:	The technical basis descriptions are often inadequate. The nature/description of the change often gets modified as the MOC request is routed for review. The change that the last reviewer authorized may not be the change as described when the first reviewer signed the form.
Possible Solution:	Require that the MOC originator consult with representatives of key groups and collaboratively develop the technical basis before the MOC request is circulated. Depending upon the nature of the change, the quorum for this might be (a) the originator, an operations representative, and a technical representative or (b) the originator, an operations representative, and an appropriate maintenance craftsperson. Note that certain electronic MOC (eMOC) documentation/approval systems have work flow management capabilities that administer the re-approval requirements associated with modifications to the requested change. Similarly, processes for handling modifications (or deviations) can be built into paper-based systems.
Comments:	None.

Problem:	Getting the required signatures on the MOC request form is proving to be a hassle, and I am concerned about the quality of the evaluation that some reviewers are using in their decision to authorize the change. Sometimes it appears that Andy and Scott will automatically sign the form if Susan has signed it before them.
Proposed Solution:	Some organizations require (or at least provide the opportunity for) reviewers to discuss and authorize changes during periodic group meetings (e.g., part of the plant staff's morning meeting).
Comments:	While this does not ensure a collaborative approach, it at least provides the opportunity for one.

Problem:	The plant can't keep up with the current volume of MOC requests. There are too many circulating at any given time, and it is nearly impossible to keep track of who has the approval package. We have lost many of these and have had to start over.
Proposed Solution:	(1) See the proposed solution immediately above (joint review meetings). (2) Consider installing an eMOC system. Paper files do not need to be circulated. The eMOC system keeps track of pending approvals.
Comments:	Many eMOC systems have document management features that allow the attachment of supporting documentation to the MOC request. One CCPS member company reports that efficient tracking of MOC progress and not losing documentation are the biggest advantages cited by users of his facility's eMOC system.
Problem:	We have problems providing personnel with convenient access to MOC records while protecting valuable historical records. Plus, we are drowning under the volume of paperwork.
Proposed Solution:	Consider installing an eMOC system that has document management system capabilities.
Comments:	None.

Problem:	The plant has problems with maintenance work orders that request changes slipping through the system without MOC reviews/controls.
Proposed Solution:	(1) Train maintenance planners and all maintenance crafts personnel on the definitions of change and RIK. Let them know that they have a responsibility to flag potential changes for review and will be held accountable for this. (2) Provide a field on the work order form for indicating whether an MOC is required (e.g., "MOC Required: Yes/No"). (3) If the answer to item 2 above us "yes," provide a field on the work order form for indicating the MOC number. (4) Perform periodic audits of work orders to identify changes that were not processed through the MOC system. Require the responsible parties to retroactively address the changes, and use these opportunities to further educate/counsel MOC system users.
Comments:	None.
Problem:	The time required to process an MOC is too long.
Proposed Solution:	(1) Consider parallel steps for routing rather than a series of steps. (2) Consider designating a single/final approver who identifies reviewers based on what parts of the organization will be impacted by the change. Note, the reviews should be done prior to final approval of the change.
Comments:	The first solution may require multiple copies of the MOC package and may complicate tracking of the package. Alternatively, converting to an eMOC system could reduce the time requirement. The second solution requires having a very experienced person as the designated MOC review path expert.

Problem:	Forms are not being filled out correctly, and originators don't know whom to contact for assistance.
Proposed Solution:	(1) Evaluate the adequacy of the training being provided and supplement it as warranted. (2) Identify and publicize one or more knowledgeable people as point(s) of contact for education/guidance on MOC implementation.
Comments:	None.
Problem:	The MOC procedure does not provide any instruction concerning records retention – what records are to be kept and for how long?
Proposed Solution:	Clear requirements should be established addressing (a) the types of information to be retained with the approved and implemented change request and (b) the length of time that this MOC package should be retained.
Comments:	Significant regulatory issues and reduced RBPS program effectiveness could result from the failure to retain needed information for an appropriate period of time (consider, for example, the need to refer to MOC documentation when revalidating a PHA. Once such regulatory and programmatic issues have been addressed, organizations may want to seek guidance from corporate legal counsel with regard to establishing an appropriate records retention schedule.
Problem:	MOC originators don't understand which PSI needs to be updated in conjunction with an MOC.
Proposed Solution:	(1) Evaluate the adequacy of the training being provided and supplement it as warranted. (2) Include a checklist of the more commonly affected PSI on the MOC approval form to reduce the possibility that a particular type of PSI will be overlooked.
Comments:	None.

Problem:	Frequent personnel turnover results in unassigned MOC action items. The person new to his or her organizational role is not aware of items previously assigned to his or her predecessor.
Proposed Solution:	(1) Consider installing an e-MOC system. Reassignment of responsibilities for action items can be readily accomplished through such systems. (2) Many organizations have an action item database/tracking system for recommendations resulting from PHAs, incident investigations, audits, and so forth. Consider integrating MOC action items into such a system.
Comments:	None.
Problem:	Sometimes it isn't clear who has the custodial responsibility for shepherding the MOC through the approval and implementation process.
Proposed Solution:	Require that the MOC originator retain the primary responsibility for shepherding the MOC through the approval and implementation process, rather than handing it off to someone else.
Comments:	None.
Problem:	Field verification is not done correctly or on time.
Proposed Solution:	This is partly a training issue. However, there may be a need to consider who is responsible for field verification and what the procedure should be. The primary objective is to verify that the installation was implemented according to the engineering design specifications. Since the verification must be timely (prior to startup), good communication is essential. The MOC process should include an effective way to notify the responsible person when the installation is ready for field verification. It is also an easy way for the responsible person to confirm and report that the field verification is complete.
Comments:	None.

Proem:	We are experiencing problems with the handoff of new technology from Research to Manufacturing. Technology packages are not well documented.
Proposed Solution:	Integrate a simplified version of the MOC system into the R&D program. Also, consider requiring close R&D support for manufacturing operations (possibly on a 24/7 basis) until a comprehensive technology package is provided to the manufacturing group.
Comments:	None.

REFERENCES

1 Kletz, T. A., What Went Wrong? – Case Histories of Process Plant Disasters, Houston: Gulf Publishing Company, 1985.

2 Center for Chemical Process Safety, *Guidelines for Technical Management of Chemical Process Safety,* American Institute of Chemical Engineers, New York, New York, 1989.

3 *Process Safety Management of Highly Hazardous Chemicals (29 CFR 1910.119),* U.S. Occupational Safety and Health Administration, May 1992, *available at* www.osha.gov.

4 Feigenbaum, A. V., *Total Quality Control,* McGraw-Hill, Inc., New York, New York, 1983.

5 Arendt, J. S., *Resource Guide for the Process Safety Code of Management Practices.* Chemical Manufacturers Association, Inc., Washington, DC, 1990.

6 Sanders, R. E., Management of Change in Chemical Plants – Learning from Case Histories, Oxford, Butterworth-Heinemann Ltd, 1993.

7 Ian S Sutton, Management of Change, Southwestern Books, 1st ed., 1998, ISBN-13: 978-1575029856.

8 *Guidelines for Risk Based Process Safety*, AIChE Center for Chemical Process Safety, Wiley, New York, NY, 2007.

9 The American Chemistry Council's Responsible Care Management System®, American Chemistry Council, Alexandria, Virginia.

10 Accidental Release Prevention Requirements: Risk Management Programs Under Clean Air Act Section 112(r)(7), 40 CFR 68, U.S. Environmental Protection Agency, June 20, 1996, Fed. Reg. Vol. 61 [31667-31730], available at *www.epa.gov*.

11 ISO 14001 – Environmental Management System, International Organization for Standardization (ISO), Geneva, Switzerland, available at www.iso.org/iso/en/iso9000-14000/index.html.

12 OHSAS 18001 – International Occupational Health and Safety Management System, available at www.ohsas-18001-occupational-health-and-safety.com/.

13 API RP 750, *Management of Process Hazards*, American Petroleum Institute, Washington, DC, 1990.

14 Control of Major-Accident Hazards Involving Dangerous Substances, European Directive Seveso II (96/82/EC), available at http://europa.eu.int/comm/environment/seveso/.

15 Control of Major Accident Hazards Regulations (COMAH), United Kingdom Health & Safety Executive, 1999 and 2005, available at www.hse.gov.uk/comah/.

16 Organization for Economic Cooperation and Development – Guiding Principles on Chemical Accident Prevention, Preparedness, and Response, 2nd edition, 2003, Organisation for Economic Co-Operation and Development, Paris, 2003, available at www2.oecd.org/guidingprinciples/index.asp.

17 *Guidelines for Hazard Evaluation Procedures (Second Edition with Worked Examples)*, Center for Chemical Process Safety, New York, 1992.

18 *Revalidating Process Hazard Analyses*, Center for Chemical Process Safety, New York, 2001.

19 *Guidelines for Performing Pre-Startup Reviews*, AIChE Center for Chemical Process Safety, Wiley, New York, NY, 2007.

20 Center for Chemical Process Safety, *Plant Guidelines for Technical Management of Chemical Process Safety*, American Institute of Chemical Engineers, New York, New York, 1992 (and revised edition, 1995).

21 Center for Chemical Process Safety, *The Business Case for Process Safety*, 2003, http://www.aiche.org/CCPS/Publications/Print/Date.aspx

INDEX

Printed and bound by CPI Group (UK) Ltd, Croydon, CR0 4YY
16/04/2025
14658346-0001